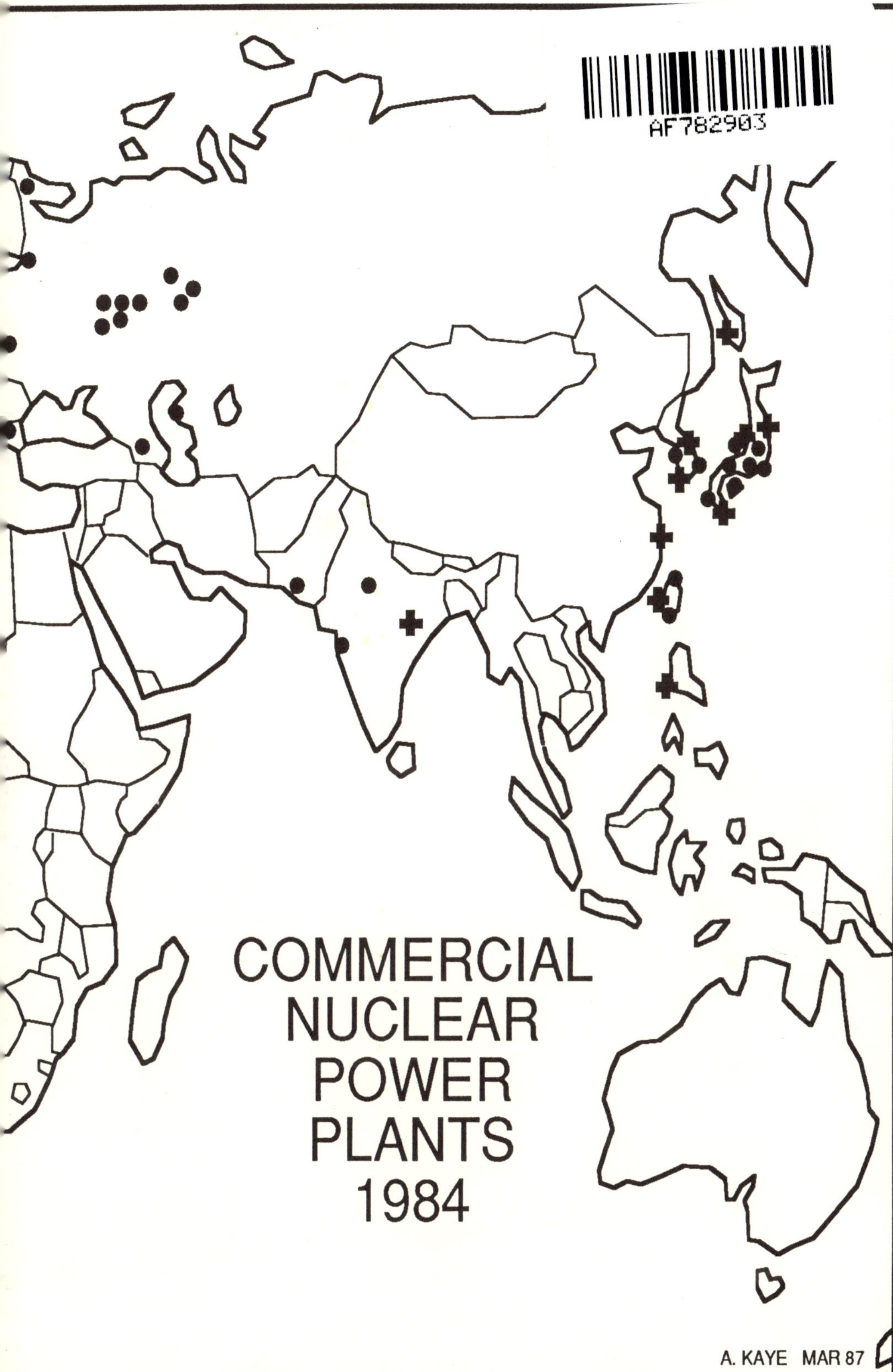

COMMERCIAL NUCLEAR POWER PLANTS 1984

A. KAYE MAR 87

How Safe?

Three Mile Island, Chernobyl and Beyond

How Safe?

Three Mile Island, Chernobyl and Beyond

JAMES MEGAW

Stoddart

Published in 1987 by
Stoddart Publishing Co. Limited
34 Lesmill Road
Toronto, Ontario
M3B 2T6

Canadian Cataloguing in Publication Data

Megaw, W. J., 1924-
How Safe: Three Mile Island, Chernobyl and beyond

Includes index.
ISBN 0-7737-2111-8

1. Nuclear energy. 2. Nuclear energy — Safety measures. 3. Nuclear facilities. 4. Nuclear facilities — Safety measures. 5. Nuclear power plants — Ukraine — Chernobyl — Accidents. I. Title.

TK9152.M44 1987 363.1'1962148 C86-094938-9

Jacket Photograph: S. Barrow, The Image Bank Canada
Jacket Design: Newton Frank
Text Illustrations: Jim Megaw and Andrew Kaye
Printed and bound in United States

Second printing October 1987

For Maureen

Contents

List of Inserts

List of Figures

List of Tables

Introduction

On 26 April 1986 an event occurred at Chernobyl which was to shake the civilised world. The biggest reactor accident ever released about 50 million curies of radioactive material which was wafted by the wind around most of Europe and even to North America and Japan. Thirty-one people were to die within the next seven weeks as the direct result of the accident and it is feared that some thousands more will contract cancer within the next fifty years as a result of the radiation dose they received from the released fission products.

In the weeks after the accident most people were understandably confused by the bewildering mass of often contradictory information given out by the Russian government, the media and the various environmental groups. It became apparent that there was a need for a dispassionate analysis not only of the accident itself and its effects, but also of the effects of radiation on people and a comparison of the social costs of the various methods of generating electrical energy.

In this book an attempt is made to explain how the accident happened, in terms the layman can understand, and what its effects will be. It offers the reader some advice on what to do in case of an accident to a nearby reactor and gives a survey of past reactor accidents. It describes the effects of radiation on the human body and examines in some detail the claims of certain experts that low doses of radiation have a greater risk than is admitted by the various official organisations such as the International Commission for Radiological Protection (ICRP). It tries to relate some commonly accepted risks of everyday life to the known risks of generating electrical energy by various methods.

An attempt is made to assess the impact of the Chernobyl accident on the future of nuclear energy.

The book was written during the weeks after the Chernobyl accident when very little information was released by the Russian authorities and, in Chapter One particularly, some rather speculative estimates were made of the radiation doses near the reactor, based on radiation measurements made in Finland, more than one thousand kilometres away. A few days after the manuscript was submitted the official Russian report became available and, rather than rewrite Chapter One with the benefit of hindsight, it was left unaltered and a new Chapter Two written giving the official measurements. In fact the somewhat crude estimates of Chapter One agree rather well with the subsequent measurements and show that it is possible to derive quite accurate figures for the close-in radiation doses from very remote measurements.

The author is indebted to a great many sources of information. Initial media coverage of the accident naturally tended to the sensational but it very quickly became more responsible. Much invaluable information was obtained from *The Times*, the *New York Times*, *Time* magazine, *The Economist*, *Atom*, the Toronto *Globe and Mail* and the *Toronto Star*. The National Radiological Protection Agencies of Denmark, Finland, Sweden, Austria and Italy were commendably rapid in producing reports of the levels of Chernobyl fallout in their countries and generous in supplying copies.

The standard works on atomic energy and nuclear power by Glasstone were often consulted and an invaluable source of information was the "Chronology of accidents at nuclear facilities, domestic and international" compiled by the *Washington Post* and obtained through *Nuclear Notepad* via InfoMedia NewsCenter.

The author is grateful to many friends and colleagues, notably Dr. J. Bromley, Dr. A. C. Chamberlain, Mr. H. Flyger, Mr. J. A. Garland and Dr. Bliss Tracey for much valuable information and useful discussion. Thanks are also due to Mrs. Sally Lakdawala and to Mr. Andrew Kaye for invaluable assistance with the diagrams. The author is greatly indebted to Dr. Archie Robertson for his painstaking reading of the final manuscript and for the very many useful suggestions he made. However, the author is entirely responsible for the opinions expressed in this book.

Finally, the book could never have been written without the patience and understanding of the author's wife Maureen who was willing even to postpone the celebration of her fortieth wedding anniversary in order that a deadline could be met!

Prelude

Valery Khodemchuk thought life was pretty good. He was lucky first of all to live in Pripyat. It was a new town, a *rabochy posyolki* (workers' settlement) thrown up rather quickly to accommodate workers at the nearby Chernobyl reactor site, but that had its advantages; you got a decent apartment if you worked at the plant. The streets were pleasant and there were plenty of flowers around. The blocks of apartments, five and ten stories high, were situated in pine groves. Each area had a school, a library, shops and sports facilities nearby. The countryside was not all that interesting, marshy sandy forestland but Lake Kijevskoje a few miles away was a vacation area and there was good fishing in the Pripyat river nearby.

As he walked through the gates at 11:30 pm on Friday 25 April Khodemchuk was already planning his weekend and ribbing his workmate Vladimir Shjashenok, (who worked on the automatic control systems) about the fish that got away the previous Sunday. Today had been a pleasant spring day with light warm winds from the south. The forecast was good for the weekend too. It was getting foggy already, but that would burn off as soon as the sun rose. In the meantime there was a heavy night ahead of them for reactor 4 was shut down and that meant lots of work.

People who didn't know any better thought that when the reactors were down everyone had an easy time. In fact, it was the other way round. The reactors shut down only for a couple of days every six months or so and that was when all the maintenance had to be done. It was do this, do that, go there, come here and there was scarcely time to catch your breath. There was an experiment going on tonight and that made things worse.

Those scientists thought they were God. Anyone would think that Chernobyl was built for their convenience and not for generating electricity.

Viktor Kibenok was going on duty at the same time. He was not a reactor operator like the others, but a fireman in the work's brigade. The trouble was that, after all the training, there weren't any fires. Not that he really wanted any, especially at Chernobyl, but it was pretty dull all the same hanging around the fire house all day waiting for an alarm that never came. Training exercise after training exercise, what to do if . . . . Anyone would think he was going to be a reactor physicist. At twenty-three he really wanted something more active. Still, perhaps it was just as well for yesterday Olga had told him that she definitely was pregnant this time. Maybe it was time he settled down anyway. Olga missed Kiev where they had lived before he got the job at Chernobyl. She said Pripyat was far too quiet but even she had to admit that the money was good and they had a much better apartment than they had had in Kiev. The baby would settle her down too.

As each of the hundreds of workers streamed into the plant none of them could have imagined that before the shift was over two of them would be dead and another twenty nine doomed. Valery Khodemchuk would be killed in an explosion and entombed forever in the reactor and Vladimir Shjashenok hit on the head by a piece of the massive 200-ton crane and able only to mutter, "there Valery" before he too died. They would never have believed either that within a few hours Viktor Kibenok would be a hero who with twenty-seven of his crew stood over the open reactor for hours trying to contain the flames and avert a catastrophe, and incurred a massive radiation dose from which he died fifteen days later. They didn't suspect that within forty-eight hours the name of their nuclear generating station would be on the lips of half the western world and that the night's events at Chernobyl would induce a chill of fear in millions of people from Poland to Canada on both sides of the Iron Curtain and, perhaps for the first time, bring home the stunning realisation that there are no frontiers in the atmosphere and that John Donne was right, four hundred years ago, when he said that "no man is an island."

Chapter One

Chernobyl: The Anatomy of a Disaster

The Beginning — in Sweden

For the rest of the world the story of Chernobyl starts, not in the Ukraine, but at a nuclear generating station, Forsmark, about 100 kilometres north of Stockholm. A worker, arriving for his morning shift on Monday 28 April 1986 walked past a radiation detector and set off its alarm. The plant operators, assuming that the high readings were due to a leak of radioactive material from their own reactor, closed it down and evacuated the plant. Subsequent tests showed that the radioactive materials did not come from Forsmark or any other Swedish reactor.

The Swedes quickly alerted the Americans who at first thought that a Russian underground weapons test must have leaked to the atmosphere. By the afternoon the Swedish scientists had identified radioactive krypton, xenon, iodine, caesium and cobalt in the fallout in proportions which indicated that they must have come from a reactor and not a weapons test. The radioactive products formed in fission all have different half lives (Insert 1) and hence decay at different rates. In a weapons test they are all formed at the same instant and start to decay at the same time. In a reactor they are formed over a long period and thus start to decay at different times, so that the proportions of the various fission products present are quite different from the weapons case. The Swedish scientists concluded that at least a partial meltdown of a reactor must have occurred somewhere and also that the radioactive cloud had started to drift over their coast at about 2 pm on Sunday. From the known wind speeds and directions they were able to work back and trace the passage of the cloud over Latvia and Minsk to Kiev.

Insert 1

Radioactive Decay

Radioactive atoms are just unstable atoms continually trying to change themselves into a more stable configuration. The greater the degree of instability the faster they change. The rate of decay is measured by the "half life" which is the time for half of the atoms in a particular sample to decay. The shorter the half life, the more radioactive the isotope. Thus, if you take a million atoms of tritium, you will only have 500,000 left after 12.5 years while, of a million atoms of iodine 131, half will have decayed after eight days. The half lives of radioactive isotopes vary from billionths of a second to billions of years. The amazing thing is that all samples of a particular isotope have exactly the same half life and no one has ever been able to change that half life or stop a radioactive isotope from decaying. The interesting question arising is "how does a particular atom in a sample know when to decay so that with all the other atoms there the half life comes out exactly right?"

Remember that half lives apply only to large numbers of atoms. If you start off with a million atoms of iodine 131 you will have exactly half a million left after eight days. If you start off with two atoms you cannot predict when there will be one atom left. It might be after ten seconds or it might take a month for the first atom to decay. Consider the parallel of life insurance; an insurance company makes money by predicting the "half lives" of large numbers of people. If it insures a million twenty-year-old men it can forecast rather well (from actuarial tables) how long it will be before half of them have died, and it can adjust its premiums so that it will make a profit. It doesn't know when an individual policy holder will die. So it is with radioactive atoms, the future of a single atom is uncertain, but that of large numbers of atoms can be predicted exactly.

In the meantime, Swedish diplomats were asking questions in Moscow. The Scientific Attaché, Per Olof Sjostedt asked the Soviet State Committee for the Use of Atomic Energy for an explanation, but was told that there was no information. At a reception at the Swedish Embassy in the evening the Ambassador, Torsten Orn, asked a Soviet Foreign Ministry official for information, but the official simply "took note" of the request. Then, at 9 pm the television news programme *Vremya* announced: "An accident has occurred at the Chernobyl Nuclear Power Plant and one of the reactors has been damaged. Measures are being taken to eliminate the consequences of the accident. Aid is being given to those affected. A Government Commission has been set up." The First Secretary at the Swedish Embassy was instructed from Stockholm to request additional information required to protect Swedish citizens. He contacted the Russian Foreign Ministry at 11 pm and, within fifteen minutes, was referred to the official announcement and

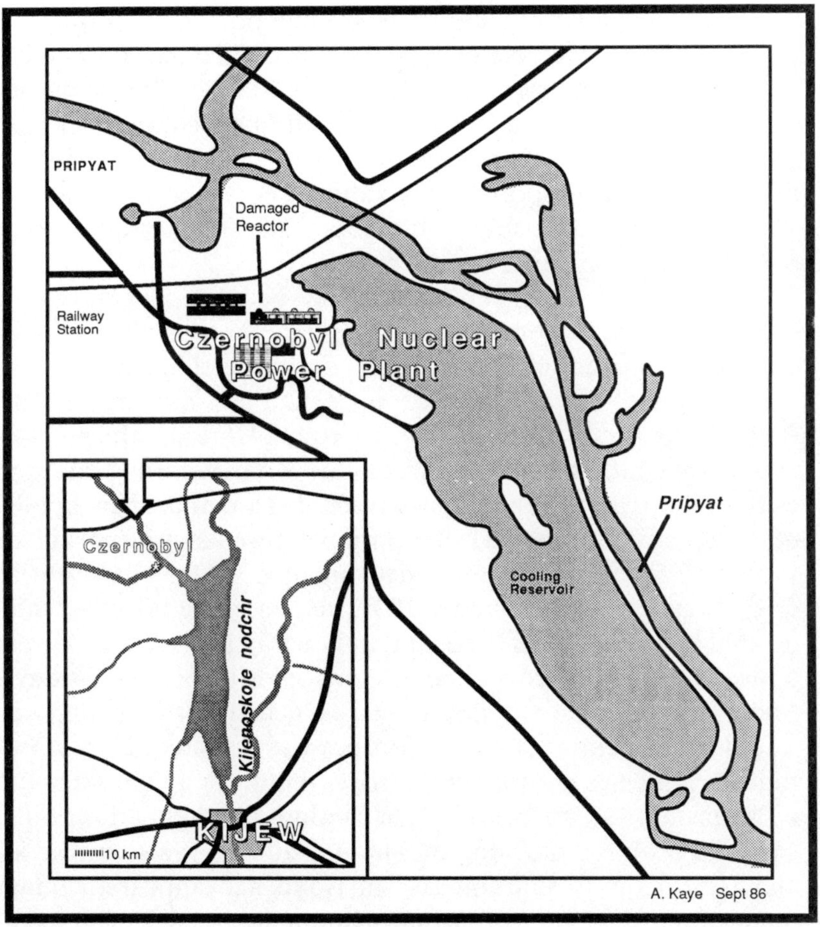

Figure 1 Situation of the Chernobyl Nuclear Power Station

told that there was "no other information to give." On 29 April (three days after the start of the accident) the International Atomic Energy Agency (IAEA) in Vienna received a short official message from the Soviet government informing it of the accident but offering no other details.

The Rest of the World

In the next few days confusion reigned in the western media with wilder and wilder estimates of the casualties and the damage done and more and more gloomy prognostications. On 30 April the Soviet News Agency, *Tass* released four short paragraphs saying that two people had been killed during the accident, added that the radiation situation had now stabilised and that four settlements near the reactor had been evacuated. By this time some reports in the west were putting the deaths at 2,000 and a West German source said that 10,000 deaths from radiation sickness could be expected.

By 30 April, other countries had become aware that a radioactive cloud from the damaged reactor was passing over them and that radioactive materials from the cloud were either being washed out by rain or deposited on the ground. In Poland, the nearest foreign country to the Ukraine, the government banned the sale of milk from cows on pasture and children were to be treated with potassium iodide to reduce their uptake of radioactive iodine 131. In parts of Sweden where rain or snow had fallen since the accident, people were advised not to drink water which might be 100 times more radioactive than normal and drugstores there and in Norway and Denmark were besieged by people seeking potassium iodide tablets. By Friday 2 May the radioactive cloud had reached Britain and an unwelcome cocktail of seven isotopes was detected. In order of radioactivity the list was iodine 131, tellurium 132, ruthenium 103, caesium 137, ruthenium 106, caesium 134 and barium 140.

At 8 am on the morning of Tuesday 29 April, a scientific attaché from the Soviet embassy in Bonn had appeared, unannounced, at the office of Atomforum, a non-government agency that represents West Germany's nuclear power industry and asked if the Germans knew anyone who could advise his country on how to put out a graphite fire in a reactor. They were advised that a variety of fairly obvious materials such as sand, water, foam could be used, but that the exact circumstances would have to be known. It was also suggested that the Russians should contact the British who had had some experience with a graphite fire at Windscale in 1957. The Russians sought advice from the Swedish nuclear authority on the same day.

The American government offered a variety of assistance including the use of a computer system that predicts the path of radioactivity in the atmosphere, a helicopter-borne system to measure and map the deposition of radioactive materials on the ground, a team of health physicists and others to examine air, water and soil, medical experts on radiation exposure and technical experts on decontamination. The list was cabled to the US Embassy in Moscow but the offer was politely declined by the Russian government which said that it had the means to deal with the situation. Moscow did, however, invite Dr. Robert Gale, a bone-marrow transplant specialist from the University of California, to come to Russia to provide medical aid to Chernobyl victims.

By this time satellite photographs showing the damaged Chernobyl reactor with a plume of smoke being emitted from it had appeared in the western press and it was realised that the graphite moderator — a large structure, 12 metres in diameter and 8 metres high, whose purpose is to slow down the neutrons emitted in the fission of uranium 235 atoms so that they can cause additional fissions in other uranium 235 atoms — was on fire.

The Accident

The graphite at Chernobyl normally operates at a temperature of 700°C, a bright-red heat. It is pierced by 1660 zirconium alloy tubes which contain the fuel and the cooling water. Within each tube is a cluster of 18 zirconium alloy cans containing the uranium oxide fuel. The cooling water is pumped into the bottom of each tube and flows up between the fuel elements, which are continually heated by the fission of uranium atoms, and the outer wall of the tube. At the top the water boils and the high pressure "wet" steam (i.e., a mixture of steam and droplets of boiling water) which results is piped first to a drying chamber (high-velocity water droplets would damage the turbine blades and have to be removed) and then to the turbines (Figure 2). The fuel element cans are arrayed within the graphite moderator rather like sausages on a barbeque and, if they get too hot, they will split, just like the sausages. However as long as the cooling

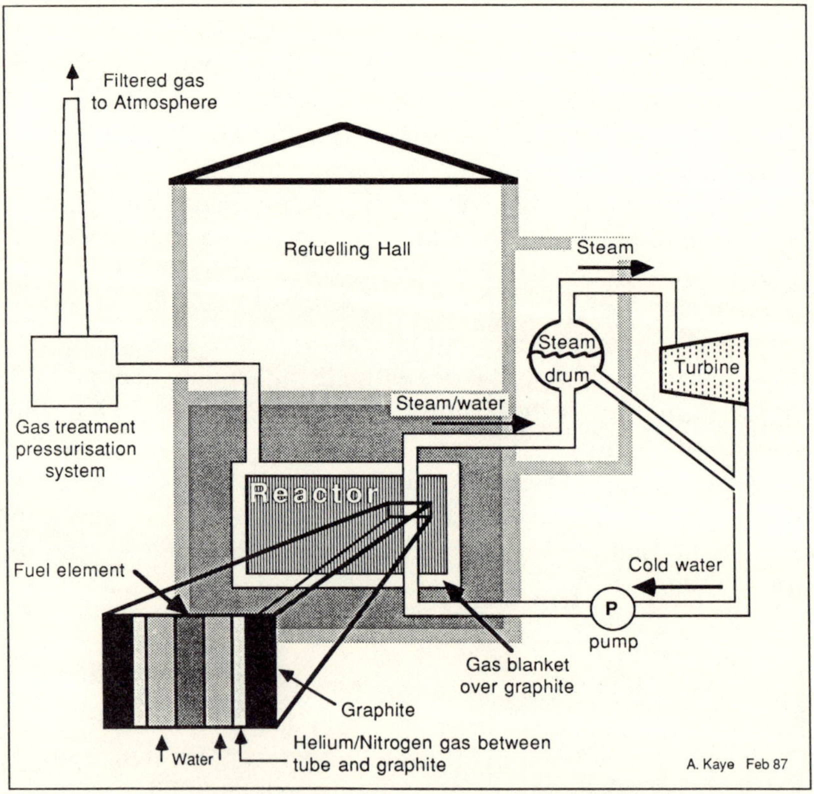

Figure 2 Sectionalised Diagram of the Chernobyl RBMK-1000 reactor.

water flows this will not happen. If there is an interruption in the coolant flow and a fuel element overheats and leaks, there is a danger that volatile fission products contained in the cans will be released into the reactor.

The first sign of things going wrong at Chernobyl was probably the triggering of a device that detects a leak of radioactive material from a fuel element can. It was almost certainly a massive leak caused by the failure of coolant to some part of the reactor core. How the initial condition occurred may never be known outside Russia, but it was announced by the authorities as being due to "human error." This suggests that some experimental procedure went wrong. At the time of the accident the reactor was working at low power and some experiments were being carried out.

Mr. Viktor Sidorenko, deputy chairman of the state committee for nuclear inspection, discussed the accident with several western journalists on 21 May. He said that there had been an explosion in the reactor but would not offer a detailed explanation of what had caused it. He did say that the reactor's output had been lowered to six or seven per cent of capacity "for planned annual repairs." "We planned to do some experiments when the reactor was on this level" Mr. Sidorenko said "The accident took place in the stage of experimental work." He would not describe the experiments or say what relation they might have had to the accident. He said only that they were connected with the "checking up of some of the systems of the station."

Ivan Yemelyanov, deputy director of the research institute that designed the reactor, said that the reactor surged from its standby level to 50 per cent of full power in just ten seconds. It is possible that an engineer caused control rods to be withdrawn too rapidly and that a sudden increase of power and local overheating in some part of the core resulted which caused failure of a coolant pipe.

Another suggestion is that a fire of unknown origin broke out in the turbine hall (Figure 2), adjacent to the reactor hall, which may have interrupted electrical supply to the circulating pumps or ruptured pipes allowing leakage of water. In either case overheating and eventual failure of one or more fuel element cans and the release of some fission products from the fuel would occur. Deprived of cooling the fuel in the ruptured can would continue to heat and, perhaps melt (the melting point of uranium oxide is 2500°C) and a large proportion of the more volatile fission products would be released into the reactor.

None of this should have happened. Every reactor has built-in automatic devices to prevent any of these events occurring. It should be impossible to withdraw the control rods at anything other than a safe rate; the failure of a cooling tube should be immediately detected, the reactor shut down and automatic emergency cooling instituted. The leakage of fresh fission products into the reactor circuit should also be automatically detected, evaluated and the reactor shut down if necessary.

The operation of any of these systems would have prevented the Chernobyl accident. That they did not implies that one or more of the systems, universal in other reactors in the world, either did not exist, was inoperative due to mechanical or electrical failure or was deliberately overridden by the operating personnel. None of these events, serious though they were to the reactor operators, should have had any effect on the outside world, or even presented any danger to the operating staff or threatened the integrity of the reactor. It was then that a potentially disastrous defect in the design of the Chernobyl type of reactors became apparent.

With local coolant failure and continued heating, whether fission or by fission-product decay, the temperature of some of the graphite must have risen well above its normal operating temperature to perhaps 1000°C. At this temperature steam from the ruptured cooling pipe playing on the hot graphite produces water gas — a mixture of hydrogen and carbon monoxide often used as a fuel fifty years ago and produced then by blowing steam through red-hot coke. While water gas is flammable, it is not in itself explosive. However, the build-up of pressure within the moderator may have allowed water gas to escape into the refuelling hall above the reactor and there mix with oxygen and be exploded by a stray spark. An alternative cause of the explosion may have been the fact that when the temperature of zirconium gets above about 870°C it oxidises extremely rapidly in contact with water or steam and liberates large quantities of hydrogen. If mixed with oxygen, the hydrogen could have exploded and perhaps blown through the reinforced concrete roof of the reactor vault.

However it happened, an explosion did occur and was sufficient to cause the damage shown in Figure 3. It was this explosion which caused the two deaths initially reported. The real facts may never become generally known, for the conclusions of Soviet commissions of enquiry are not usually made public. While there is a moral obligation on the part of the Russian government to make the fullest disclosure to the IAEA, it cannot be enforced.

The Weather

As the deadly brew of mixed fission products was expelled from the reactor it was carried aloft by the heat released from the fire. The weather that night was typical of a high-pressure situation.

Figure 3 The Damaged Reactor

Winds at ground level were very weak and their direction was variable, and a vast area of fog developed during the night. Because of the stable atmosphere and the heat of the fire, the fission products were pushed high into the atmosphere and here the wind pattern was more definite. At a height of 1500 m the wind speeds were in the range of 8–10 m/sec and were from the east at Chernobyl, although farther away they veered towards southeast.

The wind direction for the night of 25/26 April at a height of 1500 m is shown in Figure 4 which indicates clearly how the plume got to Scandinavia. By midnight on 26 April the wind

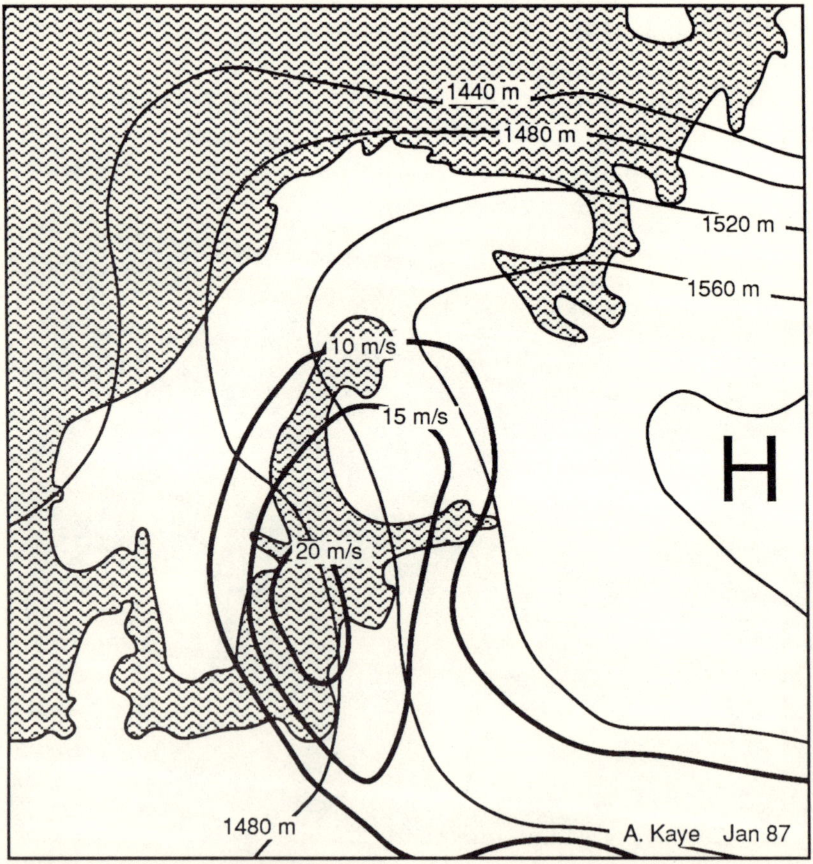

Figure 4 Meteorological Situation Over Europe on 26 April 1986. The "isobars" show the height (metres) of the 850 millibar pressure levels

had backed to easterly and was carrying the materials emitted at that time towards western Europe and this pattern continued until noon on 28 April when a westerly wind started to blow and carried any freshly released material farther into Russia. The winds then veered towards the west and north and until 1 May, when it was announced that the release had ceased, the fission products were carried in a generally southerly direction. The spread of the fission-product cloud during the days following is shown in Figure 5.

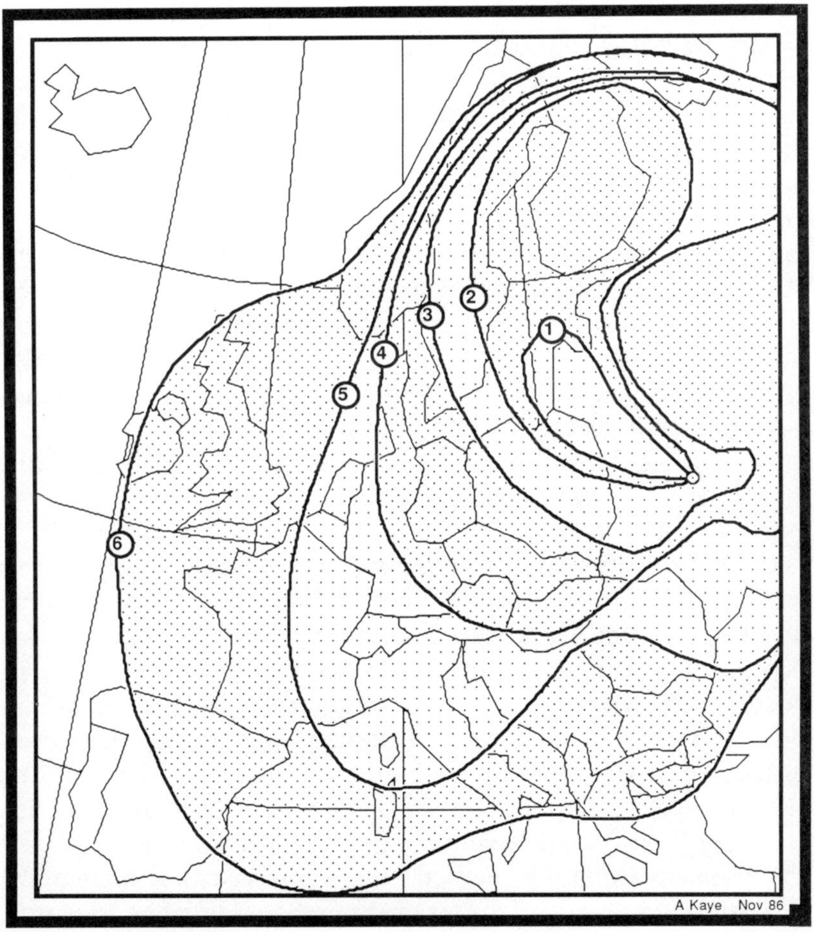

Figure 5 Dispersion of the Fission Product Plume from Chernobyl at OOGMT 1, 27/4; 2, 28/4; 3, 29/4; 4, 30/4; 5, 1/5; 6, 2/5. Courtesy ENEA-DISP, Rome

Remedial Actions

In order to stop the release of fission products, it was essential to reduce the temperature as quickly as possible. Various means of extinguishing this fire were tried and it was undoubtedly during this period of conventional fire fighting that a number of firemen incurred heavy radiation exposures from which they have since died. Eventually helicopters were used to dump sand, lead and boron onto the reactor.

The reasons for this choice of materials are not clear. The purpose of the sand was presumably to prevent air from getting to the red-hot graphite and promoting further oxidation. Lead is primarily a shielding material and it may be that the Russian engineers hoped that it would melt and flow around the core of the reactor to provide additional radiation shielding. The use of boron (a neutron absorber), implies that the engineers were uncertain as to whether the chain reaction had stopped or thought that there was a danger of its starting up again spontaneously. Such dumping of any materials in an uncontrolled way is fraught with danger for, even if the fire, as such, is extinguished, the shut-down core of the reactor still generates a great deal of heat from its radioactive decay and needs cooling. Sealing off the reactor by dumping sand, lead and boron onto it must increase the risk of further melting of fuel. The fact that it was done at all indicates the degree of desperation felt by the Russian engineers.

By 1 May the Russian government announced that the fire was out and dismissed western reports that thousands of people had been injured. It repeated the announcement that only two people had been killed in the accident and 197 admitted to hospital. Of these, 49 were discharged after medical examination. It was also reported that about 90,000 people had been evacuated from their homes in the immediate vicinity of the reactor. At a press conference on 6 May Soviet officials, including Mr. Boris Shcherbina, a deputy prime minister who led the team investigating the accident, explained that a chemical explosion occurred in the early hours of Saturday 26 April, causing a fire which spread to one of the four reactors at the site. The fire, they said, was now out and radiation levels at the

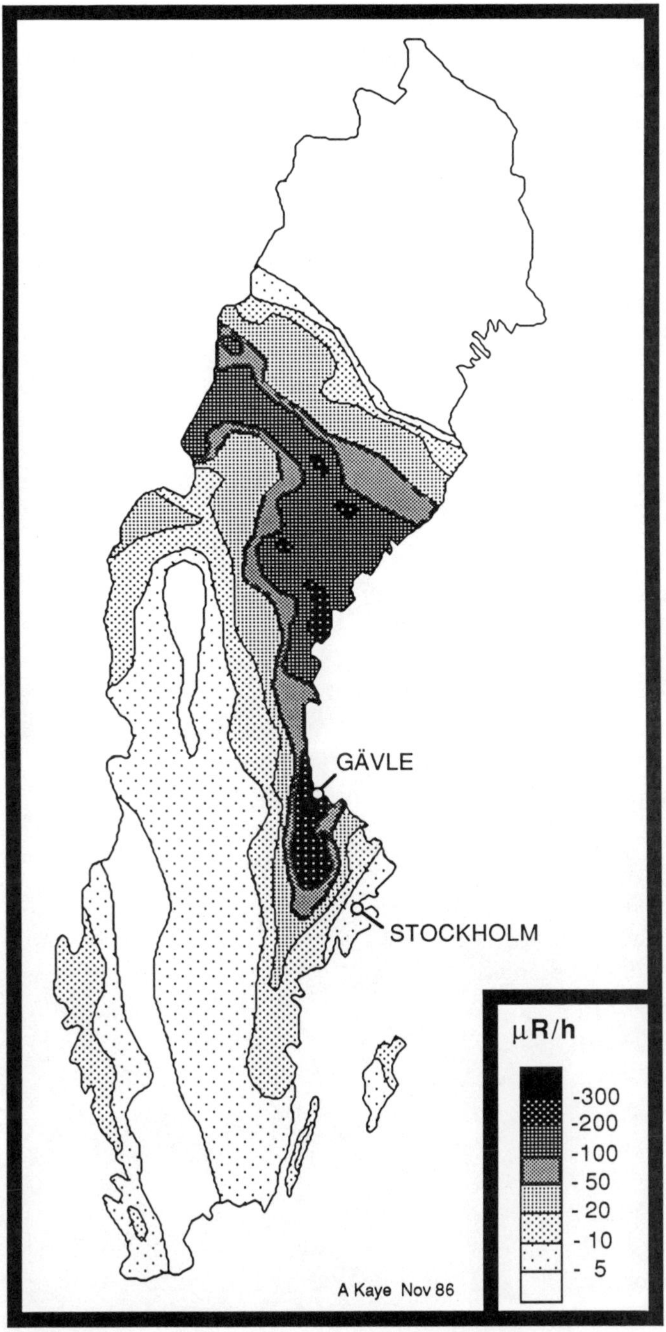

Figure 6 Radiation Dose in Sweden Due to Chernobyl (Courtesy National Institute of Radiation Protection, Stockholm)

site had dropped considerably. Mr. Shcherbina put the blame for the confusion which followed the accident on the officials at Chernobyl who at first failed to report the full extent of the danger. News of the explosion is said not to have reached Moscow until the evening of Sunday 27 April, by which time monitoring equipment in Sweden, about a thousand kilometres away from the accident, was already registering sharply increased radiation levels.

The IAEA Inspection

A three-member team from the IAEA arrived at Chernobyl on 5 May. The IAEA, a United Nations Agency, exists to pool information from all countries interested in the peaceful development of nuclear energy in order to advise and help individual states if necessary. The Agency does not exercise any power over its members and does not monitor individual nuclear power programmes, although it has produced extensive safety guidelines for almost every type of reactor. However, no country is bound by them. On 9 May the IAEA team issued a report which is given below.

Report of the IAEA Inspection Team 9 May 1986

A full and authoritative description of the accident, the reasons for it, and consequences of it can only be given by the Soviet authorities after necessary analysis.

We have, however, prepared the following brief description which is based upon what we have learned from a number of talks with ministers, officials concerned and experts, and, to some extent, from our own visual observations during a helicopter flight around the Chernobyl nuclear power plant. Many of the facts given here have been made public earlier.

- On 26 April at 1.23 am explosions occurred in the fourth unit of the Chernobyl nuclear power station. At that time the reactor, which was undergoing a planned maintenance shutdown, was at 7 per cent power level.

- As a result of the explosions, two persons lost their lives, one from hot-steam burns, the other of injuries from falling objects.

- So far, there are only hypotheses regarding specific reasons for the accident. Research and detailed analyses are under way. Records of data from the control room were recovered after the accident and are being studied.

- The explosions caused fire. The reactor building, equipment in it, the reactor itself and the reactor core were extensively damaged, resulting in radioactive releases beyond the nuclear power station area.

- The chain reaction automatically stopped at the time of the accident. This is confirmed by the fact that medical examination of persons affected showed no evidence of high neutron flux exposure.

- A large part of the radioactive releases consisted of the emission of short-lived radionuclides. Measurements indicated that up to 50 per cent of the releases were in the form of iodine 131.

- Fire teams came quickly to the site. Work was complicated by the fact that neither water nor chemicals could be used. Firemen and some nuclear power station personnel were among those injured by radiation.

- Most residents in adjacent areas were indoors at the time of the accident, thus reducing their exposure.

- In the early morning of 26 April, monitoring equipment registered increased radioactivity and reported this.

- Evacuation began on 27 April, starting with women and children. Up to 48,000 people were evacuated from Chernobyl and other locations within a 30-kilometre (19-mile) radius.

- As a preventive measure, potassium iodide tablets were widely distributed inside as well as outside the 30-kilometre zone.

- 204 persons, including nuclear power station personnel and firefighters, were affected by radiation from first degree to fourth degree, 18 persons being in the fourth degree*. All 204

---

*A Russian classification of radiation sickness. Fourth degree is the most severe, with symptoms developing within half an hour. See also pp. 59, 60.

persons were hospitalised in Moscow and treated medically. In some cases, bone-marrow transplants were performed.

- The radioactive releases from the damaged unit have been significantly reduced by shielding and neutron-absorbing material — sand, boron, clay, dolomite and lead — dropped from helicopters over the reactor resulting in decreasing radioactivity levels in the 30-kilometre zone.

- Necessary shift personnel work on the site to keep the undamaged reactors in safe shutdown condition. The third reactor, adjacent to the fourth, was not damaged in the accident and its safety systems for cooling are functioning.

- The damaged reactor suffered some fire in parts of the graphite. These fires have been extinguished, but temperatures remain high. Recriticality is not considered a problem. The aim is to encase the whole of the fourth unit in concrete and work has begun to place a concrete foundation under the reactor.

Although no systematic data on radiation levels were made available, some values were given. Maximum radiation levels within the 30-kilometre zone were 10 to 15 millirems an hour (see Insert 2). By 5 May it had decreased to two to three millirems an hour. On 8 May it had dropped to a maximum of 0.15 millirems an hour at the perimeter of the zone.

The level of radioactivity within Kiev's water reservoir was within normal limits at all times.

Inconsistencies in the IAEA Statement

Although the IAEA statement carries an air of authority, there are some obvious anomalies in it, especially with regard to radiation levels. It is possible, from the evidence obtained from neighbouring countries in the few days following the accident, to make some rough estimates of what the radiation dose (see Insert 2) must have been at places nearer to Chernobyl. The direct radiation dose from a source would be expected to fall off with distance as the square of the distance (i.e., the dose at a distance of 10 m would be 10 × 10 or 100 times less than the dose at 1 m from the source). At greater distances the absorption of the gamma rays by the air through which it is passing reduces the dose markedly and direct radiation effects from the source are unimportant beyond a distance of a kilometre or so.

Insert 2

Radiation Dose and the Measurement of Radioactivity

How do you measure the quantity of radioactivity in a substance? The obvious method is to weigh it, but this isn't good enough, for a gram of iodine 131 (half a gram of which will decay in 8 days) will emit much more radiation than a gram of strontium 90, whose half life is 30 years. In the years following the discovery of radioactivity much of the work was done with radium and in 1910 it was decided to give the name *curie* (abbreviation Ci) to "the quantity of radium emanation in equilibrium with one gram of radium." This definition was later extended to any radioactive material and the current definition of the curie is "the quantity of any radioactive material undergoing 37 billion disintegrations per second." A later unit, the *becquerel*, (abbreviation Bq) has since been defined as "that quantity of a radioactive substance undergoing one disintegration per second." Thus one curie is equivalent to 37 billion becquerels.

When ionising radiation passes through the body it interacts with the molecules of tissue, knocks electrons off them, and forms electrically charged pairs of ions. For this reason it is referred to as "ionising radiation." The original unit of radiation dose was the roentgen, named after the discoverer of X-rays, and defined as the quantity of X-rays or gamma rays required to produce one electrostatic unit of electric charge in a cubic centimetre of dry air. The damage done to the body by ionising radiation is directly proportional to the amount of energy liberated in it by the radiation and, in energy terms a roentgen is equivalent to the absorption of 83 ergs of energy per gram of dry air.

In the same way *rads* and *grays* measure the dose in any substance (for example, body tissue). A rad is that quantity of radiation dose corresponding to the absorption of 100 ergs of energy per gram of tissue and one gray is 100 rads. However that is not quite good enough, for different kinds of radiation have different effects on biological tissue and another unit, the roentgen equivalent man, or *rem* was invented to take account of this. One rem is that quantity of radiation dose which produces the same biological effect as one roentgen of X-rays. One *sievert* is equal to 100 rems.

The erg is a tiny quantity of radiation (it takes 42 million ergs to raise the temperature of one gram of water by 1 degree Celsius) and it seems almost incredible that the trivial quantity of energy represented by a few hundred rems is enough to kill a human being within a few weeks.

If radioactive material is released into the atmosphere however, the story is quite different. As the radioactive material is released from the reactor, it is first of all diluted by the wind. If the wind speed is one metre per second, each second's worth of output from the reactor will be spread over a distance of one metre. If the wind speed is 10 metres per second the same quantity of material will be spread over 10 metres. Thus, the higher the windspeed, the more dilute the cloud. Weather is also important in other ways, for atmospheric conditions determine how quickly the radioactive cloud will spread (and thus be diluted), both laterally and vertically. Finally rain falling through the cloud will wash out some of the radioactivity. Thus the cloud will be further diluted but at the cost of contaminating the ground below it with the material washed out (Figure 7).

The radiation dose received by the people over whom the cloud passes will depend on how dilute the cloud has become and how high it is above ground level when it passes over, the more dilute and the higher the cloud, the less the radiation dose to the people below. As the radioactive cloud gets farther from the reactor the fact that it gets more and more dilute, also means that it gets bigger and bigger and it will eventually grow so large that it will sweep along the ground. At this point people on the ground will actually be *in* the cloud so that they will not only get an external radiation dose from the radioactive materials in which they are immersed, but also an internal dose from those they have inhaled with the air they breathe.

The result of these dilution effects is that the concentration of radioactivity in the air falls off at a rate rather less than the

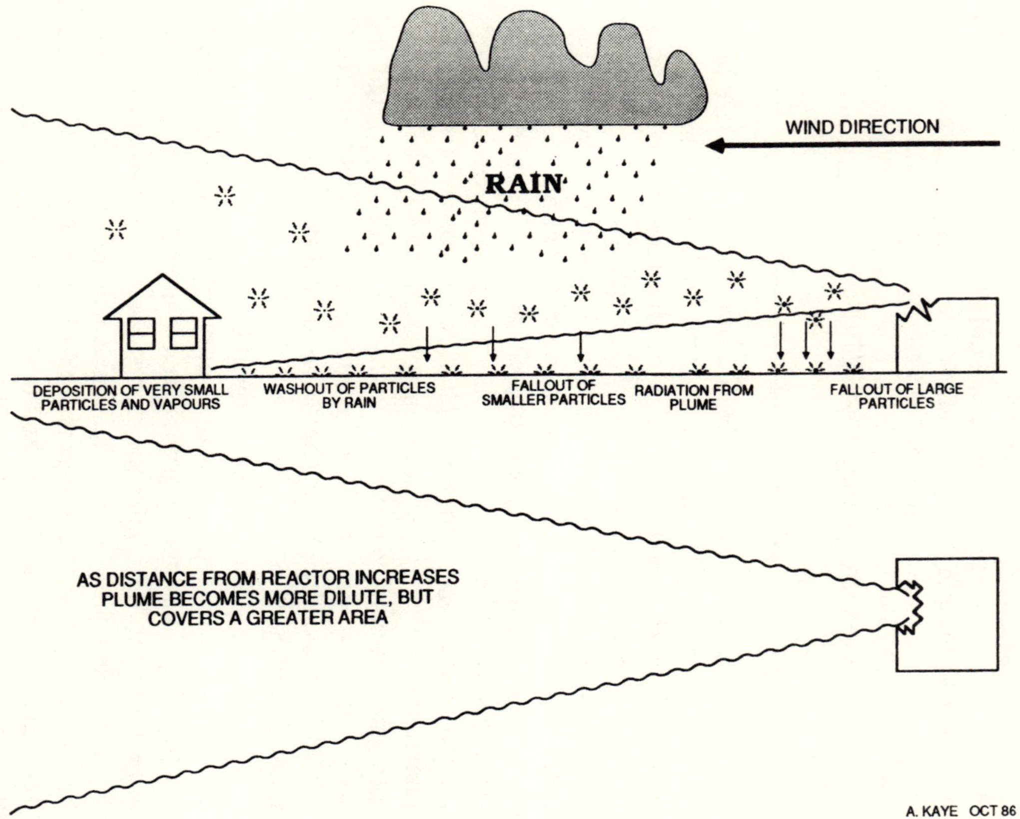

Figure 7 Idealised Behaviour of Fission Product Plume

square of the distance from the source, and it is possible, using this fact, to start off with a measured value of the radiation dose at a known distance from the reactor and extrapolate backwards to arrive at an estimate of the radiation dose rate at places closer to the reactor.

Measured Dose Rates

Initial reports of the accident (e.g., the *London Times* of 29 April) quoted the Swedish Defence Ministry as saying that the dose rate of "a few millirems per hour" recorded by several monitoring stations on the afternoon of 28 April "was not thought high enough . . . to be a danger to human beings although specialists said that the dose rate was twice as high in Finland." This prompted a remark by Ed Zebrowski, an American scientist: "If it is a few millirems an hour a thousand kilometres away I'd hate to be within ten kilometres of it." Other reports implied a rather lower dose rate in Sweden of "from 6 to 10 times the natural background." The natural background radiation is about 10 millionths of a rem per hour (see Insert 2) and this would imply an upper level of the dose rate in Sweden of about 0.1 millirem per hour. This seems a very much more credible level than that of the Defence Ministry. The maximum dose rate recorded in Finland was 0.4 millirem per hour at Uusikaupunki on the shores of the Gulf of Bothnia northwest of Helsinki (STUK, 1986).

In an important study in 1971 Jack Beattie and Pam Bryant of the UK National Radiological Protection Board give estimates of the external gamma radiation dose received from a fission-product cloud after various distances of travel. Their calculations only go up to a distance of a hundred kilometres, but a reasonable extrapolation suggests that the radiation dose received at a distance of a thousand kilometres might be 1/200,000 of that received at one kilometre from the source. One kilometre is the normal exclusion distance for nuclear power plants, i.e., there should be no places of habitation nearer the reactor than that. If the Uusikaupunki measurement was made on the centre line of the plume a calculation, admittedly rough, implies that the dose rate outside the generating station boundary must have been about $0.4 \times 200{,}000 = 80{,}000$ millirems per hour or 80 rems per hour.

This is a very high dose rate and contrasts markedly with the IAEA statement that the maximum radiation dose within the 30-km zone "has been 10 to 15 millirems per hour" and, indeed, on common sense grounds, the IAEA assertion cannot be correct. It is possible that the statement was made as the result of a misunderstanding (although one would have expected an important announcement such as this to have been thoroughly checked with the Russian authorities before release) and that the meaning was intended to be "the maximum radiation level *at the boundary* of the 30-km zone was 10 to 15 millirems per hour." This certainly makes more sense, although, if the 80-rem-per-hour estimate is accepted and a further calculation made using Beattie and Bryant's data to estimate the dose rate at 30 km, the result is 340 millirem per hour, still a factor 20 larger than the IAEA figure.

On the other hand, if it is assumed that the IAEA figure of 15 millirems per hour applies to a distance of 30 km from the reactor, then the dose rate at a distance of 1 km would have been about 4 rem per hour — still a considerable dose rate, but one which would allow evacuation even after 36 hours to take place without many cases of acute radiation sickness appearing.

Remedial Measures

Dose rates in the rems-per-hour range are very serious and, while they may not produce many acute cases of radiation sickness they do cause long-term health risks and an immediate evacuation of the people living in the area affected by the passage of the plume should undoubtedly have been ordered. This the Soviet authorities failed to do and the delay may have caused casualties which otherwise would have been avoided. The evacuation did not occur for about 36 hours when orders were received to evacuate an area 10 km in radius around the reactor. Between 2 pm and 4.20 pm on Sunday 27 April, 25,000 people were evacuated from the town of Pripyat, 4.8 kilometres from the reactor and altogether 49,000 people were removed from this area. On 3 May the exclusion area was enlarged to a 30-kilometre radius around the reactor and the number of people evacuated rose to 92,000. If the 80-rem-per-hour estimate is correct many people must have been exposed, before evacuation, to radiation doses which will certainly result in cases of radiation sickness, some of which may well be fatal (Table 1).

It is likely that the reactor staff were so pre-occupied in dealing with the disaster at their feet that they failed to appreciate that the local population was in danger. That this could have happened implies a failure of the local site-emergency plan. In all emergency organisation there should be a pre-arranged and pre-practised plan in which the site emergency controller has the responsibility for liaising with the civil authority and advising on the need for evacuation or other precautions. It is a cardinal principle that the site emergency organisation come instantly into being on a pre-arranged alarm and is a separate organisa-

Table 1

Effects of Whole-body Exposure to Radiation

Dose (in rems)	**Short-term Effects**
0–50	No detectable effects.
50–100	Temporary depression of white blood cell count. Possible nausea at upper end of exposure range.
100–200	Nausea, diarrhea, possible loss of hair, temporary sterility, not immediately fatal but long-term risk of cancer.
400–500	Acute nausea and diarrhea within a few hours. 50% of victims will die in about four weeks after showing signs of apparent recovery.
600–1000	Heavily depressed white blood cell count, acute nausea and diarrhea, 80 to 100% of those exposed will die in four weeks after sometimes showing signs of recovery.
1000–5000	Virtually 100% chance of death, sometimes within a few hours.

tion divorced from the operational staff who, at that moment, should be thinking of nothing but dealing with the physical emergency itself. Thus the normal operational staff are free to deal with the emergency and do not have to worry about its effects on the rest of the reactor site and the outside world.

The Site Emergency Controller should have a staff of engineers and health physicists who monitor the accident by taking radiation dose measurements, measuring the radioactivity of air samples both inside and outside the site at places which have been prearranged in the emergency plan. He bases his decisions on these measurements, the wind speed and direction, the sectors affected and where, in the affected sectors, there are people living. Site emergency personnel are usually made up of reactor staff who happen to be off duty at the time of the accident but who are on call, are well trained in emergency procedures and are thoroughly practised in the emergency plan. The indications are that there was no such organisation at Chernobyl, or that it failed to operate effectively.

When the decision to evacuate people was eventually made it appears to have been carried out with efficiency and despatch. To evacuate 25,000 people (there are some reports that farm animals were also evacuated) in two-and-a-half hours is no mean feat. That a further evacuation had to be ordered a week later, on 3 May, illustrates another human failing in dealing with the consequences of reactor accidents, the tendency to err on the side of conservatism and take the minimum possible action, presumably to avoid disrupting the lives of more people than necessary. It is a serious mistake to do so, for when a further evacuation has to be ordered the local population naturally forms the impression that the authorities do not know what they are doing. It is undoubtedly better to do too much rather than too little and to evacuate more people than seems strictly necessary at the time and then to allow some of them to return home after a few days.

The Victims

The two first casualties of Chernobyl were caused in the initial explosion, one, whose body was never recovered, in the explosion itself and the other "by falling objects," probably when the heavy roof crane fell. The other initial injuries were caused among the firemen fighting the fire and some of these incurred very heavy radiation exposures of several thousand rems. It was reported on 30 April that 197 people had been admitted to hospital. By 13 May *Tass* reported that six people had died of "radiation burns" and that 35 were in "grave condition." Mr. Gorbachev announced on 14 May that 299 people were in hospital and that seven had died.

On 16 May, Dr. Robert Gale, the American surgeon who went to Moscow to help with bone-marrow transplants (Insert 3) said that a total of 13 people had died and two weeks later the total had risen to 23. On 29 May Dr. Angelina Guskova, a radiologist, told the official Novostni news agency that about 80 of the Chernobyl patients under her care in Moscow's Hospital No. 6 were in extremely critical condition.

On the same day Yevgeny Chazov, a leading Soviet surgeon and deputy health minister, told reporters in Cologne that 210 people were still receiving treatment as a result of the accident. He said that all the people most seriously affected by radiation were either staff of the nuclear power plant or firemen rushed in to fight the blaze and that the 129 people most seriously affected had been taken by plane to Moscow within 12 hours. Dr. Chazov added that no cases of acute radiation sickness had been diagnosed among the approximately 100,000 people who were subsequently evacuated from the area around Chernobyl. He disclosed that the 21 people who had died from the effects of radiation included 11 who had received bone-marrow transplants (Insert 3). This was the first report that some, at least, of the attempts to save radiation victims with marrow transplants had failed.

At the same meeting, a congress of International Physicians for the Prevention of Nuclear War, Dr. Leonid Ilyin, Director of the Institute of Biophysics in Moscow described how, after the accident, the rescue team immediately identified about a thousand people who had been exposed to radiation, but only 299 were regarded as having serious pathological injuries. Of those 129 were immediately evacuated to Moscow and the

Insert 3

Bone-Marrow Transplants

The bone marrow is one of the vital regions of the human body, for it is the source of the body's blood and immune defence cells. It is also the region that is most damaged by heavy doses of radiation, for radiation hits hardest those cells which are dividing most rapidly. This is why the fastest growing cells of the body — the bone-marrow cells, hair cells and cells of the gastro intestinal tract are preferentially destroyed by radiation. Thus people who are irradiated lose their hair, vomit and lose their immune systems, the cells of which originate in the bone marrow.

The last-ditch desperate treatment for someone who has had a very high dose of radiation is therefore a bone-marrow transplant. It has succeeded in the past, for five workers involved in an accident to research reactor at Vinca in Yugoslavia in 1958, and who had received doses of up to 800 rems, were flown to Paris for treatment. Four of them received bone-marrow transplants and survived. They received bone marrow from relatives which may, or may not have matched their own.

The operation itself is not considered difficult. Bone marrow is removed from the donor by syringe, usually from the bones of the pelvis, the marrow is separated from the blood in the sample and injected into the patient's veins. Only a few cubic centimetres of marrow are required and it will, by itself, seek out the marrow spaces in the bones, where, if the transplant is successful, the new cells will repopulate the patient's marrow. Such transplants are used more or less routinely to treat cases of aplastic anaemia and leukemia. The main problem is that, to be successful, the tissues of donor and recipient must match almost perfectly. If the match is not good enough the transplanted marrow may not survive or, just as bad, may recognise the patient's tissues as foreign and attack them. It may be necessary to screen several thousand potential donors to find one who is a close enough match to the patient, and this cannot usually be done well in a hurry. The best candidates for donors are siblings, each of whom has a 25 per cent chance of being the same marrow type as the patient. Even under the best of circumstances a bone marrow transplant is a fairly desperate struggle to save the patient's life. When a large number of patients arises at once, as in the case of Chernobyl, the odds are heavily against getting good enough matches in time and this may be one of the reasons for the apparently high rate of failure of the Chernobyl transplants as opposed to the Yugoslavian case.

Another is undoubtedly that many of the Chernobyl patients suffered much higher radiation doses. It is estimated that between 100 and 200 persons received more than 300 rems and from 400 to 500 people received more than 100 rem. Of 13 patients who received ordinary bone marrow transplants 5 were still alive after 10 weeks and were out of immediate danger. There can, however, be major complications 3 months after a transplant.

---

remainder flown there by special planes the next day. Dr. Ilyin said that the most serious cases were those who had received doses of 2,000 rems or more, which leads to breakdown of the central nervous system. These victims were lying prostrate and unable to move. It is probable that these were the "eighteen persons being in the fourth degree" mentioned in the IAEA report. The next category had incurred exposures of from 200 to 1,500 rems. They suffered vomiting, sometimes continuously, but after the initial shock, they went into a period of apparent recovery and were in a state of euphoria by the time they were transported to Moscow. Before long, however, they were again feeling very ill. Dr. Ilyin added that doctors knew in advance that bone-marrow transplants had little hope of success. "We had to try it for humanitarian reasons," he said, "but it didn't work. I never thought it would." He felt that more of the victims would die soon although he could not give an exact forecast.

By 5 June the total number of deaths had risen to 26 and Oleg Shchepin, a first deputy health minister, said that half of those who died had received bone-marrow transplants. Dr. Gale confirmed the number of dead as 26 on 6 June and said that he felt that "the figure will probably not change by more than 10 per cent." This was the first intimation that the death toll from acute radiation exposure was levelling off.

Those victims who died in Moscow are buried in a special plot in the cemetery at Mitino, a quiet village half an hour's drive from central Moscow. On 3 August it was reported that two fresh graves, heaped with flowers, had appeared there. On the graves are cards saying "Novik, A. V., July 29, 1986" and "Vershinin, Y. A., July 23, 1986." On 6 August another grave appeared, that of K. I. Luzganova, apparently the first female casualty of the accident, with a card saying that she had died on 4 August. None of these names has appeared in official announcements although many of the names on the other graves have been mentioned in press reports of heroism at Chernobyl.

Long-term Effects

The likely long-term effects of the Chernobyl accident on the world will be discussed in Chapter Five. In the Soviet Union the results have already been considerable. Quite apart from the casualties and disruption of the everyday life of the evacuees, the economic effect on the country as a whole has been serious. According to the *New Scientist* (1 May 1986), the Soviet Union now has 28 of the Chernobyl (type RBMK-1000) reactors operating. Other sources say 15, but there is general agreement that the RMBK-1000 reactors account for half of the nuclear generating capacity in the USSR.

Initial reports say that all of these have been shut down in the aftermath of the accident which means that nearly 30,000 megawatts or six per cent of the country's entire generating capacity will have to be replaced from other sources or done without. Even if the USSR has the spare generating capacity (which it is thought not to have since too many of its old and inefficient coal- and oil-fired stations are still operating because the new equipment which was to have replaced them was not ready on time) the extra cost of alternative sources will be considerable. This disruption will be temporary, for there are already reports that even the remaining reactors on the Chernobyl site will be running at full power by the end of 1986. Whether this happens will, of course, depend on the findings of the commission of enquiry into the causes of the accident. If it is established, for example, that the cause of the accident was operator error and not some design failure, there would be no reason to delay the restart of the reactors.

Insert 4

Inside Hospital Six

A few days after the accident the central drama of the Chernobyl disaster switched to an austere brick building in a northwest suburb of Moscow. Hospital Number 6 had years ago been chosen as one of the hospitals which would deal with the victims of a nuclear accident or nuclear war and the rather limited resources available were concentrated here. Here too were brought the worst cases from Chernobyl.

Patient A has received three times the LD 50* but is, amazingly, a month after the accident, still alive. He is watched over by three nurses and kept inside a plastic tent, his only defence against infections, for his immune system has been destroyed by the radiation. Even through the hazy plastic it is possible to see the effects of the radiation — the loss of hair, the blotchy, mahogany-coloured skin and the burns. Only the medical team and his visiting relatives know his identity. They must don gowns and sterilised masks to talk to him. Amazingly, he is lucid and willing to talk about the accident.

Five minutes after the explosion occurred he knew he was being irradiated, but he stayed inside the damaged reactor hall trying to stop a meltdown of the core. He worked for two hours and then began to vomit. He kept on for another two hours. Asked why he did it, he replied, "It was my job; besides my wife and children live near Chernobyl and another explosion would have been a catastrophe." Patient A, a 55-year-old mechanical engineer, has not been told whether he will survive, perhaps because nobody knows. The odds were heavily against him from the start, that he is alive now is a minor miracle, and the passage of each day adds a little hope to the medical team.

Among the other patients is Leonid Telyatnikov, whose two sons wait at a camp for Chernobyl refugees to hear of their father's fate. Another fireman is Alexander Petrovsky who says he is feeling fine. "I want to go home as soon as possible" he tells the doctors.

London Sunday Times, 29 June 1986

*the radiation does which is lethal to 50% of the population exposed to it.

Long-term Health Effects

On 6 June it was announced that Dr. Gale had signed a long-term agreement with Soviet health experts to follow the health of about 100,000 people who were felt to be at highest risk due to the accident. He predicted that the number of people who would eventually contract cancer as a result of the accident would be "rather small" and that the disease would develop over several years, making it difficult to link the cases directly to the Chernobyl accident. He said that, while estimates of the long-term consequences should be regarded as preliminary, it was reasonably clear that estimates of tens of thousands to hundreds of thousands of additional cases of cancer were very likely to represent substantial overestimates. This is a reassuring statement, but it is rather vague. A more detailed discussion of the effects of low levels of radiation is given in Chapter Four.

Long-term Agricultural Effects

An accurate estimate of the long-term agricultural effect of the Chernobyl accident would require more definite estimates of the total radioactive release than are presently available, including the proportions of the various isotopes and of their deposition behaviour.

Estimates of the release vary between one and three per cent of the reactor's inventory of fission products and H. J. Dunster (Director of Britain's National Radiological Protection Board) has suggested that the release "was at least 100 times worse than that at Windscale." The accident at Windscale, England in October 1957 released 27,000 curies of iodine 131, 600 curies of caesium 137, 200 curies of strontium 89 and 2 curies of strontium 90 to the atmosphere, and if Dunster's estimate is correct, about three million curies of iodine 131, 60,000 curies of caesium 137, 20,000 curies of strontium 89 and 200 curies of strontium 90 might have been released at Chernobyl.*

---

*Subsequent Russian estimates give a total of 50 million curies (excluding noble gases), measured on 6 May, released.

A US estimate (from Lawrence Livermore National Laboratory, quoted in *Science*, 232, p. 814, 16 May 1986) suggested that at least 50% of the iodine and caesium (representing 30 to 40 million curies of iodine 131 and 3 million curies of caesium 137) escaped from the reactor. While this latter estimate is being used by the US Task Force on the Soviet Nuclear Accident set up on 29 April, it seems likely to be an overestimate. After the direct radiation dose received from the cloud passing overhead, it is the iodine 131 which causes the most immediate threat to health. However it is a threat that is relatively easily resolved by stopping the distribution of milk from cows which have grazed on affected pasture. Since it takes between two and three days for the iodine to work its way through the chain — deposition on grass, concentration in cow's milk, consumption by babies and small children — there is time to deal with the situation and there is no reason to doubt that the Soviet local authority did that. The iodine 131 is also a short-term problem since its radioactive half life is only 8 days, so after a period of three months the level has been reduced by a factor 2435, and a factor of six million after another three months. A great deal of concern was expressed that supplies of drinking water would be contaminated due to the proximity of the River Dnieper which is used as a source of drinking water for the city of Kiev. At Windscale it was found that the possible intake of radioiodine from drinking water was very small compared with the potential intake from contaminated milk. Samples of eggs, vegetables and meat from the more highly contaminated areas around Windscale were also shown to be insignificant as a source of iodine 131 in the human diet relative to milk.

When radioactive materials are deposited on soil (as opposed to on the leaves of growing crops) the proportional uptake to the crops via the roots depends on a number of factors, particularly the amount of replaceable calcium and the proportion of clay in the soil. Caesium 137, the second-most-abundant isotope released at Windscale has a half life of 30 years and might thus be expected to be a long-term threat to agriculture if it were deposited on soil and eventually incorporated in successive crops via the root system.

In fact, according to the British Medical Research Council

(1960), caesium 137 is held in the soil in forms which are largely inaccessible to plants and its uptake into crops is determined by direct deposition on to the leaves of the plant. In soil with a high clay content the caesium is tightly bound to the soil and rather little is taken up by plants. If the soil has a high humus (organic) content, rather more caesium will be absorbed by the plants. It can be expected therefore that damage to agriculture as a result of deposition of caesium 137 will only occur to the crops that were growing at the time of the accident and will not constitute a threat to next year's crops. However there are exceptions and there may be a higher uptake of caesium through the roots of grass in natural grazing land which is not fertilised. The direct radiation to agricultural workers from caesium deposited on the ground may also give unacceptably high radiation doses to people working the land in future years in areas where the deposition was very heavy and safe working will probably entail scraping off the top few inches of topsoil and burying it safely elsewhere.

Strontium 90 has a half life of 28 years and would be deposited both on the ground and on the leaves of plants. It will be taken up by plants from the soil through the root system together with the chemically similar element calcium. The plant does not differentiate between the two provided that they are in the same chemical form, so that the freely soluble calcium in the soil will act to dilute the strontium as far as uptake by the plant is concerned. Thus a high calcium content in soil results in a low strontium uptake by plants. Strontium 90 which is deposited on plants will be absorbed through the leaves and this will not be associated with calcium, so that no dilution will take place. The amount of strontium deposited on the leaves which is absorbed varies with the type of plant and is usually considerable in grasses. Uptake of strontium 90 through the root systems of plants is usually of minor significance compared by the direct uptake by the parts of the plants which are above ground.

It would seem therefore that the major effects on agriculture in the area around Chernobyl from iodine 131, caesium 137 and strontium 90 will be largely limited to this year's crops and will only have a minor impact on the agricultural future of the

Ukraine. The Ukraine grows about 40 per cent of the Soviet Union's winter wheat, but that is mainly to the south of Chernobyl and, to start with at any rate, the radioactive plume went north. American traders spoke of a possible 10 per cent loss in Russian's grain harvest this year, but that is probably an overestimate. Even if it were not, there is a large surplus of grain in world markets. Most of Russia's big sugar beet growing areas are to the south and east of Chernobyl, again in the opposite direction to that of the initial plume. The Ukraine produces 60 per cent of Russia's 8.5 million tonne sugar crop, but even if the entire Ukraine crop were wiped out Russian buying would fall short of the world surplus of between 8 and 15 million tonnes.

It is in the supply of milk that the most significant Soviet rise in short-term demand is likely. World supplies of milk and butter are also plentiful. In 1984 the Soviet Union imported nearly 60,000 tonnes of milk powder; the EEC has 700,000 tonnes of milk powder in stock and the butter mountain contains 1.3 million tonnes.

It thus seems probable that the effects of Chernobyl on Soviet agriculture will be slight and limited in the main to this season's crops. Any shortfall can easily be met on the world markets without any great increase in prices. Wheat prices in Chicago certainly overreacted and at times the changes in prices reached the limits imposed by the exchange. The long-term effect on agriculture will be negligible. According to the criteria employed by the UK Atomic Energy Authority, up to 150 square miles of farmland may have been affected badly enough to render bread, milk and vegetables from them dangerous. Only about ten per cent of this would remain dangerous after a year. There is apparently no intention of cleaning up or of salvaging the reactor itself. Reports suggest that it is intended to seal the reactor in a concrete block, provided with cooling pipes to take away the fission-product heat.*

*A Reuter report on 17 November 1986 stated that the sealing of the reactor had been completed.

The Aftermath

By the middle of June Chernobyl had disappeared from the front pages of the world's newspapers and any items that did surface were tucked away on inside pages. On 19 July there was a Politburo announcement on television news which said that it had been established that the accident had been caused by a series of gross breaches of the reactor operational regulations by workers of the atomic power station. It added that experiments with turbogenerator operation regimes were conducted at the fourth generating unit when it was sidelined for planned repairs at night. "The managers and specialists of the atomic power station had not prepared for that experiment, nor approved it with appropriate organisations, although it had been their duty to do so. Finally, proper supervision was not organised when those experiments were carried out, nor proper safety measures taken." The report concluded: "The Ministry of Power Engineering and Electrification and the State Atomic Power Inspection were guilty of lack of control over the situation at the Chernobyl Plant and did not take sufficient measures to secure compliance with safety regulations and breaches of discipline and operational regulations." The report announced the dismissal of three high-ranking officials and the initiation of criminal proceedings against individuals responsible for gross breaches of discipline and operational regulations at the power plant. It did not identify those against whom criminal proceedings were in progress but those dismissed included G. A. Shasharin, a Deputy Minister of Power and Electrification, Aleksandr G. Meshkov, a First Deputy Director of Medium Machine Building and the deputy director of the Research Centre where the Chernobyl reactors were designed, Ivan Y. Yemelyanov. It had been announced the day before that Y. V. Kulov, director of the three-year-old State Committee for Safety in the Atomic Power Industry had been relieved of his duties, and, a couple of weeks previously that the Director of the Chernobyl station and his Chief Engineer had been dismissed. The statement also said that the Electric Power Minister Anatoly I. Mayorets "deserved to be relieved of his duties but had received only a reprimand because of his short tenure in the office."

The Politburo announced the formation of a new agency, The Ministry of Atomic Power Engineering, "to raise the standards of management and responsibility for the development of nuclear power engineering." It said that it had instructed ministries and departments to carry out a series of measures to ensure the safe operation of existing nuclear power plants, including tightening of discipline and a new programme to retrain and rate nuclear power plant workers, and improved inspection procedures. The report stated that direct economic losses as a result of the accident were about two billion roubles (US$2.7 billion) and that the loss of electric power from the closed plant "had created difficulties with the power supply to the national economy."

Lessons to be Learned

The Chernobyl accident has already aroused feelings of what can only be described as holy glee among the opponents of nuclear energy. "We're seeing the beginning of the end of the nuclear industry" said Mr. Jonathon Porritt of Friends of the Earth. The fact that Soviet spokesmen had boasted in the past that the Chernobyl-type reactors were "so safe that the odds of a meltdown are one in ten thousand years" is advanced as an argument that all reactors are unsafe and should be closed down forthwith. Our perception of risk is so flawed that we fail to appreciate that these odds, while they may still be correct, do not mean that a meltdown will not occur for ten thousand years, but that it may occur anytime within that period, whether tomorrow, next week or in five thousand years time.

What are the lessons to be learned from the Chernobyl disaster? The first is surely that there is no way to conceal an accident to a reactor which allows any significant quantities of radioactive materials to escape to the environment. The Chernobyl accident was detected at a distance of more than a thousand kilometres. And not only detected, the finger of suspicion was pointed inexorably at the source.

The natural tendency of the operators seems to be to try to allay the fears of the general public. What they should be trying to do is to inform the public honestly and to the best of their ability what has happened and what they are doing to correct the situation. Even before any accidents they should be trying to inform the public of the types of accidents that can happen, the precautions being taken to ensure that they don't happen, and the remedial action that will be taken if they do.

The questions and objections raised by action groups such as Friends of the Earth, Greenpeace and Energy Probe should be treated seriously and answered honestly and patiently. The action groups serve a useful purpose for they ask questions, sometimes awkward ones. If they didn't already exist, they would have to be invented. The fact that many of the questions they ask are stupid shouldn't faze the reactor operators. Stupid questions are the easiest to answer. Of course some of their questions are clever and rhetorical and the questioner doesn't really want an answer. Those are the really annoying ones, but they have to be lived with.

The next lesson that must be learned is that most accidents happen when the reactor is shut down or near shut-down, for that is when changes are made and experiments installed. There is a tendency for operators to breathe a sigh of relief and relax when the reactor reduces power. The two biggest releases of fission products from reactors, Windscale and Chernobyl occurred in, or just after, shut-down conditions. Reactors running at full power are practically trouble free and therein lies another source of danger. It is very difficult to keep always on the alert when nothing ever happens, yet that is what operators must do. They must be highly trained, highly professional — and always prepared for trouble. There are indications that the operators at Three Mile Island did not achieve this high standard of readiness and the suspicion must be that the Chernobyl operators also failed to do so.

Next, every reactor must have a highly efficient and highly trained site emergency organisation ready to jump in and take over the moment an emergency occurs. The local civil authority must work closely with this organisation and take part in its exercises. Training of the Site Emergency Teams must be continuous and effective so that all of their members respond instantly to the emergency warning and each knows exactly what to do. The conception of ever more fanciful exercises is a difficult task for the reactor management, but it is a vital one; there have been so few reactor accidents in the history of nuclear power that the next one is always bound to be different from the last.

If these lessons are learned there is hope that the next accident, for there *will* be a next one, will be controlled and contained safely and that the mistakes which occurred, not only at Chernobyl, but also at Three Mile Island and at Windscale, will not be repeated. It may be that, in a hundred year's time Chernobyl may be seen as the beginning of the end of the nuclear era. If so, it will have been as the result of a political rather than a technical decision. Chernobyl may persuade many people that nuclear energy is too expensive at any price. Few people yet understand the price of doing without it.

Chapter Two

What Really Happened at Chernobyl

In mid-August 1986 the Russian State Committee for using the Atomic Energy of the USSR produced a 382-page report for the IAEA Expert Conference on the Chernobyl Accident which was held in Vienna from 25 to 29 August. The report gives a remarkably frank account of the shortcomings in design and reactor operation which led to the accident. The immediate cause was an experiment which it had been planned to perform while the reactor was being shut down for a "medium repair" on 25 April. The purpose of the experiment was to determine whether, after steam had been shut off from the turbogenerator, the inertia of the still-rotating generator would be sufficient to generate enough electricity to operate auxiliary motors which were part of the emergency cooling system of the reactor.

In a previous similar test on the same reactor it had been found that the voltage from the generator dropped sharply long before the generator stopped rotating. In the new test it was planned to use a special system to regulate the magnetic field of the generator to try to eliminate this shortcoming. The report says that the programme for this test was not prepared or approved in the proper way and was of low quality, the section on safety measures was composed only "as a matter of form" and the plan involved the disconnection of the emergency cooling system of the reactor throughout the period of the tests (about four hours). In addition "personnel deviated from carrying out the programme thereby creating the conditions for the development of an emergency situation."

The reactor was operated normally until 1 am on 25 April when the operators began to reduce reactor power in preparation for the test. Turbogenerator number 7 was disconnected from the grid at 1.05 pm when the reactor output was 1600Mw (thermal). At this point the electrical supply for the four main cooling pumps and two feed water pumps (which were normally powered from generator number 7) was transferred to generator number 8, the second of the two turbogenerators on this reactor. The emergency cooling system was disconnected from the large diameter pipe lines at 2 pm in accordance with the test plan. However the further running down of the reactor was delayed because of a request from the grid distribution centre at Kiev. The operation of the reactor *without* its emergency cooling system continued, in gross breach of the operating regulations, for the rest of the day.

The test eventually began at 11.10 pm. The programme stipulated that the test was to be carried out at a reactor power of 700–1000 Mw (thermal). To enable the reactor to operate at this low power, the local automatic control system had been disengaged. This caused an imbalance of the automatic regulator which the operator was unable to correct (manually) quickly enough. As a result the reactor power dropped to below 30 Mw (thermal).

At this point many of the neutrons needed for fission were being absorbed by the fission product xenon and it became difficult to sustain a chain reaction. The reactor power could be raised only by progressively withdrawing control rods and it took the operators until 1 am to get it back to 200 Mw. This was much below the power level specified for the test (700Mw), but it was decided, nevertheless, to proceed.

By this time both the automatic regulator rods (inserted and withdrawn from the bottom of the reactor) and the control rods (inserted from the top of the reactor) had been withdrawn almost to their fullest extent. At 1.03 and 1.07 am, two additional cooling pumps were switched on, in addition to the six pumps which were operating. This was to ensure that at the end of the experiment, (for which four of the pumps were required), four pumps would remain operating to provide reliable cooling of the reactor core. The operation of these pumps greatly increased the flow of cold water into the reactor, decreased the rate of steam-generation and reduced the level of water in the steam separator below the emergency level. This would normally have shut the reactor down and, to prevent that from happening, the operators short-circuited the emergency protection signals.

Meanwhile the reactivity of the reactor continued to drop slowly and at 1.22.30 am the operator noticed that it had dropped to the point at which regulations required that the reactor be shut down. At this time the "reactivity reserve" was the equivalent of a total of 6 to 8 (inserted) control rods as opposed to the 15 to 30 required for stable operation. Nevertheless the test proceeded.

At 1.23.04 am the shut-down control valves of turbogenerator number 8 were closed and the reactor continued to operate at a power of 200 Mw (thermal). The emergency control of the valves of turbogenerator number 7 had been disconnected during the previous afternoon, so that the test could be repeated if the first attempt was unsuccessful. This was a departure from the test programme.

At this point the condition of the reactor was that it was operating at 200 Mw (thermal) and still generating steam. There was nowhere for the steam to go since both turbogenerators were shut off. The steam pressure thus began to rise and at the same time the flow of water through the reactor began to decrease since four of the eight pumps were working off the generator which was deprived of steam and running down. Further steam-generation resulted and, since steam absorbs neutrons much less effectively than water, there was a rapid rise of reactivity and the reactor power increased to more than 530 Mw (thermal) in less than three seconds. At this point (1.23.40 am) the operator pushed the emergency "SCRAM" button which inserted all control and shutdown rods into the reactor.

By this time the automatic control rods were at the bottom of the core (where they had been withdrawn in order to get the reactor power up to 200 Mw) and the periodic manual adjustments made by the operators in their attempt to achieve the conditions for the test had resulted in practically all the other control rods having been withdrawn to the top of the core. When the SCRAM button was pressed the rods went down but the operator saw that they stopped before they reached the correct depth. He then cut off the drive couplings so that they fell into the core under their own weight.

Personnel who were outside the reactor building reported that they heard two explosions, one after the other, at 1.24 am. Red-hot fragments and sparks flew up above the reactor building and some of them fell on the roof of the turbogenerator room and started a fire. A computer analysis of the probable course of events leading to the explosions suggests that the continuing decrease of the water flow through the channels under the conditions of increasing power led to intense steam-formation in the fuel channels, a breakdown in the cooling of fuel elements, rapid heating of the fuel and its disintegration, the boiling of whatever cooling water remained and a steam explosion which led to the destruction of the reactor and part of the reactor building and the eventual release of fission products to the atmosphere.

The sharp temperature increase in the reactor core, the rupture of the cooling channels (releasing steam on to the red-hot graphite moderator, producing water gas) and the chemical reaction between the overheated zirconium canning and water — (releasing hydrogen) ignited by the "fireworks of flying hot and glowing fragments" produced by the steam explosion — resulted in the second explosion.

The inevitability of the headlong descent to disaster is illustrated in Table 2 which lists the successive mistakes perpetrated by the operators, their reasons for making them and the consequences to the reactor.

Dealing with the Emergency

On the night of 25/26 April there were 176 duty operating personnel and other workers in the various shops and repair services on the site of the first and second phase of the Chernobyl station (these two phases contained four operating reactors). In addition there were 268 construction workers on the night shift at the site of the third phase where two additional reactors were being built.

As a result of the two explosions red-hot and flaming

Table 2

Deliberate Actions by the Reactor Operators Leading to the Accident

Action	Motivation	Results
Withdrawing too many control rods too far.	To overcome the effects of neutron poisoning* by xenon and increase reactor power to the planned level for the tests.	Impossible to shut down reactor quickly.
Power allowed to dip below level specified for tests.	Operator error in disengagement of local automatic regulator.	Reactor goes to unstable, difficult-to-control state.
Running all pumps at more than permissible flows.	To fulfil requirements of test programme.	Loss of cooling.
Blocking system to shut down reactor when both turbogenerators are disengaged.	So that test could be repeated if necessary.	Loss of automatic shut-down capacity.
Blocking system to protect against low water levels in separator.	To attempt to conduct tests despite unstable condition of reactor.	Loss of automatic cooling capability.
Disengagement of emergency cooling system.	To prevent reactor shutting down during test.	Loss of possibility of reducing scale of accident.

*the fission product xenon is an avid absorber of neutrons.

fragments of the reactor core were ejected on to the roofs of nearby buildings and more than thirty fires were started. Intense fires were started in the turbine room above turbogenerator number 7, in the reactor room itself and in the partially destroyed compartments next to it. It apparently took an hour and a half for firefighting units to arrive from Pripyat (just outside the three-kilometre exclusion zone) and from Chernobyl (15 km away). The initial firefighting efforts were directed to preventing the spread of the fire to the nearby reactor number 3.

The report states laconically: "By two hours and five minutes (after the accident) most of the fires had been put out on the roof of the turbine room and by two hours 30 minutes on the roof of the reactor building. By 05.00 the fire had been put out."

This must rank as the most low-key reporting ever of a disaster situation that can probably best be described as a hell on earth. The firemen were working in unimaginable conditions of searing heat, their boots sticking in the molten bitumen on the roof, and, although they may not have known it initially, they were being exposed to a potentially lethal radiation dose. It cannot have been long before the first of them succumbed to the earliest effects of radiation sickness, and before they became aware of that extra dimension of horror in their situation. That they carried on regardless of their personal fate amply justifies their description as heroes in the Russian press. No description is given of measures taken to limit their radiation dose, but it was during this phase that the highest radiation doses must have been received.

The initial classification of their exposure was based on how quickly the individual became sick. The term "Extremely severe" or "fourth degree" was applied to those individuals who exhibited symptoms (vomiting, headache, rise of temperature) within half an hour of exposure. There were 22 people in this category and they developed radiation burns to between 40% and 90% of their body surface. They were estimated to have received whole-body doses of between 1200 and 1600 rads. One of them died on the fourth day after the accident and another 18 between day 10 and day 50. The third-degree acute radiation sickness victims developed their initial symptoms between one half and one hour after exposure and were estimated to have received whole-body doses of between 420 and 630 rads. There were 23 of these and seven of them died between two and seven weeks after the accident. Of these, six had severe burns which greatly aggravated their condition.

Second-degree victims showed their first reaction between one and two hours after exposure and were estimated to have received doses of between 200 and 400 rads. There were 53 of these, and very few had skin burns. The first-degree victims did not show their first symptoms until two hours or more after exposure and were thought to have had a radiation dose of between 80 and 210 rads. The report lists only the deaths occurring during the 50 days after the accident and accounts for 26 of the 29 deaths known to have occurred from the effects of radiation. The three deaths not mentioned are presumably of those victims who are mentioned on pp. 43–44 as having died on 23 and 29 July and 4 August. The extent of their radiation exposure is not known. No deaths occurred among patients with first- and second-degree radiation sickness and at the time of writing of the report thirty patients were still in hospital.

Radiation Dose in the Neighbourhood of the Reactor

From the beginning of the accident and during the fires which followed the wind carried radioactive materials over the town of Pripyat. Later on the plume actually enveloped parts of the town and radioactive materials were deposited on the ground. From 9 pm on 26 April the gamma dose rate measured at 1 metre above ground level in different streets of the town varied between 14 and 140 millirad per hour. The radiation situation in the town worsened later on and at 7 am on 27 April the gamma dose rate in the region of the town nearest the power plant (Kurchatova Street) was from 180 to 600 millirad per hour. This trend continued and by 5 pm on 27 April (i.e., after the evacuation, which started at 2 pm, had been completed), the dose rate in the Kurchatova Street area was from 720 to 1000 millirad per hour. The dose rate dropped only slowly during the next three days. The report states that the population of Pripyat was warned to stay indoors immediately after the accident started and that all open air activities were banned on 26 April.

It is estimated that people inside houses were exposed to 2 to 5 times less radiation dose than they would have obtained outside and for this reason it is considered that the probable dose levels for the population of Pripyat before it was evacuated was between 1.5 and 5 rads gamma and from 10 to 20 rads beta dose to the skin. Measurements of the iodine 131 content of the thyroid gland of 206 inhabitants after they were evacuated indicated that the thyroid dose to 200 of them was less than 30 rad. No information is given as to the dose received by the other six.

Radiation Dose to the Rest of Russia

Considerable effort was devoted to determining the integrated population dose for the whole of the affected area of Russia. The calculated integrated dose for the year 1986 is 8.6 million man rem. and that for the next 50 years is 29 million man rem. The annual collective dose received in the same period by the population from natural background radiation is 10 million man rem.

The resulting calculated increase in the mortality rate from cancer among this population over the next 70 years from the external radiation dose is estimated in the report at 4,500 cases as compared to 9.5 million cases expected naturally. Similarly the increased mortality from thyroid cancer might be 1500 cases in the next thirty years as compared to the 150,000 expected to occur naturally. It should be stressed again that these effects are calculated from a linear extrapolation of effects received at much higher dose rates and that this procedure is considered by the International Commission on Radiological Protection (the generally accepted world authority on the effects of radiation on the human body) to result in an upper limit to the effects.

The report considers the significance of caesium 134 and 137 incorporated into foodstuffs. At the time of writing the report caesium had been incorporated into food only as a result of deposition on grass etc from the air and incorporation in plants from the soil was insignificant. It is forecast that the 1986 grain and potato harvest would be relatively clean. However future problems are expected in the Poleys'ye region around Chernobyl for the soil there is rather poor and has a low humus content which is expected to result in a caesium uptake ten, or even a hundred, times greater than in other soils. The integrated dose from deposited caesium incorporated into food for the population in the Ukraine and Byelorussia is estimated to be 2.1×10^8 man rems which may result in an increased mortality rate "not exceeding 0.4%" or 40,000 additional deaths over the next 70 years.

These estimates were discussed in great detail at the Vienna meeting and the general conclusion among western radiation experts was that they were probably overestimates. Dr. Morris Rosen, Director of the IAEA Division of Nuclear Safety arrived at an estimate of a total of about 25,000 extra cases of cancer over the next 70 years and Dr. Beninson, Chairman of the International Commission on Radiological Protection, said he considered the Soviet figures "an extreme overestimate." Two days later Dr. Beninson said that the total number of excess deaths might be as low as 5,100 and Dr. Rosen suggested 10,000 as the most probable figure. Each said that they had revised their original estimates downwards because they believed that the calculated doses, particularly due to caesium 137 may have been overestimated.

The report also considers possible future casualties among the evacuated people and estimates that their total collective exposure is 1.6 million man rems. It adds "taking into account the fact that spontaneous deaths from cancer over a 70-year period for the 114,000* evacuated people may be about 14,000 cases, the natural death rate from cancer among the exposed population will be increased by less than 2% as a result of the accidental release at Chernobyl." This implies an upper limit of 280 additional deaths from cancer among the evacuated population.

The Quantity of Fission Products Released

The report gives estimates (with a possible error of 50%) of the quantities of fission products released from the reactor. Twelve *million* curies are thought to have been emitted on 26 April

*In other parts of the report, and in press reports, the total number of people evacuated is given as 135,000.

(Figure 8) and, although the rate of release fell sharply, a further 14 million curies were emitted by 2 May. The daily rate of release then began to rise again, so that 4, 5, 7 and 8 million curies were released up to 5 May and then declined to a release of 0.1 million curies on 6 May. The senses quickly become dulled by large numbers and one tends to forget that the seemingly trivial release of 100,000 curies on 6 May was about 3000 times the quantity of iodine 131 which escaped during the Three Mile Island accident. The total quantity of radioactive material released (excluding the rare gases) was thus about 50 million curies (measured at 6 May) which represents about 3.5% of the total quantity of radionuclides in the reactor.

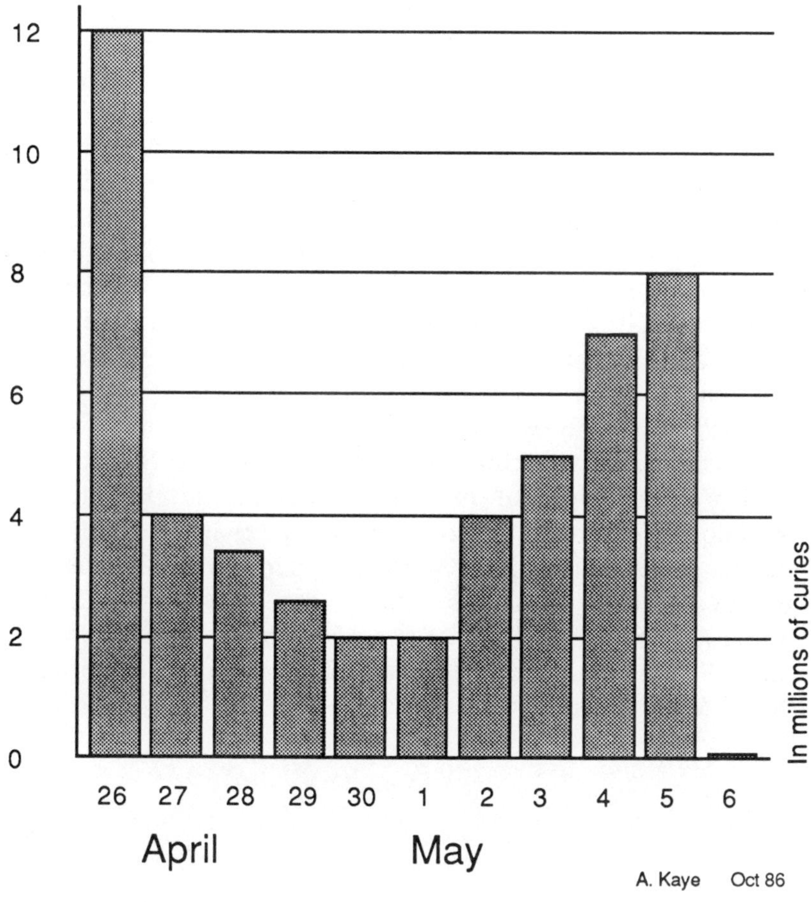

Figure 8 Emissions of Fission Products from Chernobyl

All estimates for the quantity of radioactive materials released are corrected for radioactive decay to 6 May. The figures thus underestimate the quantities released before that date.

The reason for the sharp rise in the rate of release of fission products after 1 May is of some interest, especially as it was on 1 May that the Russian government announced that the fire was out! It may have appeared so, for by this time firefighting crews had dumped several thousand tons of wet sand, boron and lead onto the reactor. The report explains the rise in the rate of release of fission products by saying that the temperature of the core rose again to 2000°C "because of residual heat." This mass of material dumped on the reactor may have, to some extent, reduced the escape of fission products to the atmosphere up to that date, but it also probably acted as a good thermal insulator, preventing the escape of heat. The heat evolved from the radioactive decay of the fission products would thus have raised the temperature of the reactor fuel even higher and resulted in the release of still more fission products. It is entirely possible therefore that the dumping of materials onto the reactor in an attempt to quell the fires, may actually have resulted in practically doubling the eventual quantity of radioactive materials released, a classic case, surely, of the cure being worse than the disease.

The release figures also, inadvertently perhaps, disclose some confusion in the original Russian report to the IAEA (p. 31) which gives values of maximum radiation levels within the 30-kilometre zone as 10 to 15 millirem per hour (the accident report gives levels of more than 1 rad per hour in Pripyat). The IAEA statement says that "On the 8th May it (the maximum radiation level) had decreased to a maximum of 0.15 mrem per hour." With admitted releases of 25 million curies between 2 and 5 May, this statement cannot be true.

Discussion of the Report at the IAEA Conference

More than 500 scientists from 50 countries assembled at Vienna on 25 August for the presentation and discussion of the Russian report on the accident. The official Russian line was to blame the operators for the accident and there can scarcely have been any report in history which laid the blame for a disaster more firmly on the people concerned. There were wholesale dismissals, demotions and promises of criminal prosecution.

This view was not entirely accepted by the scientists from other nations who agreed that the immediate cause was the actions of the operators, but nevertheless argued strongly that a prime factor had also been features of its design which made the reactor a difficult one to control. The most important of these was probably the fact that when large voids, or steam bubbles, are allowed to form in the cooling water, some of the neutrons which would normally have been absorbed by the water are made available to cause further fissions and a large and uncontrollable power surge occurs. This "positive void coefficient" was undoubtedly the single factor that transformed successive blunders of the Soviet operators into a catastrophe.

Lord Marshall (Chairman of the British Central Electricity Generating Board) said that his government had warned the Soviet Union nine years before that the RBMK-type reactors had serious defects. "You can make any design safe by having clever enough operators," he said "but the designers of Chernobyl gave the operators too difficult a task." Lord Marshall said that the Soviet Union chose this design because it would save money, and that the Russians knew it had defects, thought they could compensate, but could not. He added: "We are not talking about hindsight here. This was a judgement made in advance." The Head of the Soviet Delegation, Professor Valery Legasov said he did not know of any British warnings in 1977. He declined to compare the quality of the Chernobyl reactors with western ones and said that the accident was caused by a series of awkward and silly mistakes by the operators. Earlier at the official meeting he had said that the Russians had realised the design problems "later than other specialists, which is a great fault on our part."

The Russians gave a list of modifications which it is proposed to make to all the RBMK reactors in order to make them safer to operate and to obviate the possibility of a recurrence of the Chernobyl accident. It is intended to increase the enrichment of the fuel to 2.4% in an attempt to make the reactors less liable to power excursions due to the formation of steam bubbles. This will make the fuel cycle more expensive and make the reactors less economical to run.

Other modifications include lengthening of the control rods and increasing their number, boosting the capacity of the cooling pumps and improving the monitoring and shutdown systems. There appears to be no intention of constructing containment vessels around the existing reactors — a very expensive undertaking which might well, however, have prevented any release of fission products at all at Chernobyl. All modern commercial reactors in the western world are enclosed in very strong "containment vessels" made from either steel or prestressed concrete which are designed to contain the effects of what the designers and regulatory bodies believe to be the "maximum credible accident."

Perhaps the most bizarre Russian comment on the accident appeared in *Pravda* on 18 September when it was alleged that the Chernobyl station was designed 'by experts in hydroelectric plant construction who were unfamiliar with nuclear reactors.' This, of course, cannot be true. The evidence is rather that the reactor was designed by experts in nuclear reactors who were forced by circumstances into designing an economical reactor which could be operated safely by an expert crew, but which nevertheless was difficult to control in certain modes of operation. In spite of this there would have been no accident had the Chernobyl operators, in their eagerness to complete the experiment, not contravened their operating regulations time and time again.

Chapter Three

Self-help in a Nuclear Accident

You live near a nuclear generating station. It has probably not made any impact at all on your life so far. However at the back of your mind must lurk the thought "What will happen if there is an accident? Can Chernobyl happen here? What will I do if it does? Is there anything I can do to save myself and my family from harm?"

In normal operation nuclear reactors are the least polluting way of generating electrical energy. They do not emit smoke, carbon dioxide, sulphur dioxide, aldehydes or, indeed, any of the other pollutants, including radium, routinely released by even the cleanest and most modern coal-fired generating stations. The core of the reactor does, however, contain prodigious quantities of radioactive materials but the radiation from these is absorbed by the thick concrete walls surrounding the reactor vessel. The radiation dose received by a member of the general public who spent all his life standing at the boundary fence of the nuclear generating station would not be significantly greater than if the reactor were not there.

The radioactive materials are the products of the fission of uranium 235. Two atoms of fission products are formed from each atom of uranium 235 which undergoes fission. They constitute the whole gamut of elements from number 30, zinc, to number 63 europium. They are almost all unstable and hence radioactive. They have half lives ranging from millionths of a second to hundreds of years. They decay by emitting gamma or beta radiation. While the reactor is working new fission products are constantly being formed, while those formed earlier are constantly decaying. Eventually the rate of formation of new fission products and the rate of decay of the old become equal, so that the quantity of fission products within the reactor remains constant at a level which depends on the power at which the reactor is operating — the higher the power the greater the inventory of fission products. They are formed within the fuel element and, ideally, stay there, contained within it, until they are removed from the reactor during a fuel change.

Some of the more volatile fission products (the radioactive noble gas, krypton 85, for example) may diffuse out of the fuel element, but will still be retained within the can. If a can should, for any reason, develop a leak, those fission products will escape from it. Such occurrences are comparatively rare (in CANDU reactors they occur in less than 0.01% of all individual fuel elements) but when bad leaks develop they are a source of embarrassment to the reactor operators, and the offending fuel elements have to be removed to avoid contaminating the primary circuit and making the detection of future leaks more difficult. Should an incident occur which causes the reactor core to overheat, numbers of cans may develop leaks or even melt off their fuel elements. Should this happen, relatively large quantities of fission products may be released into the primary circuit and even into the reactor containment vessel. Provided they stay there they are of concern only to the reactor operators. However an accident in which the containment is breached and small or large quantities of fission products are released to the environment cannot be ruled out. It has happened on three occasions in the last thirty years, in 1957 at Windscale, England when 27,000 curies of iodine 131, 200 curies of strontium 89, 2 curies of strontium 90 and 600 curies of caesium 137 were released to the atmosphere, in 1979 at Three Mile Island, Pennsylvania (between 15 and 30 curies of iodine 131 released) and in April 1986 at Chernobyl, Ukraine (release about 50 million curies). What is the effect of such releases on people living in the area surrounding the reactor and what can they do to protect themselves?

The Effects of Radiation on People

We live in a radioactive world and it has always been so. The earth is continually being bombarded by cosmic radiation from outer space, the sea contains more uranium naturally than will ever be consumed by all of the world's nuclear reactors. We could escape from the cosmic radiation by building concrete houses with thick roofs, except that building materials, derived from rock and soil, also contain radioactive materials, and we all get a radiation dose from the houses we live in. The earth continually emits radon, a radioactive gas with a string of radioactive daughter products and, as if that were not enough, our bodies are themselves radioactive and, again perfectly naturally, contain the isotopes potassium 40 (half life rather more than a billion years) and carbon 14 (half life 5,500 years). In addition we voluntarily incur other radiation exposure in the doctor's office or the dentist's chair, or even when we take a flight between Europe and North America.

The quantity of radiation dose is measured in a bewildering array of units — the roentgen, the rem, the rep, the rad, the sievert and the gray (Insert 2). All of them can be related to the quantity of energy absorbed in the body due to the radiation passing through it, and the first four are roughly the same, within twenty per cent. A rad, for example, is the quantity of radiation dose resulting from the absorption of 100 ergs of energy per gram of tissue. An erg is really a tiny quantity of energy (it takes 42 million ergs to raise the temperature of one gram of water by one degree Celsius) and it is an awesome fact that the rather trivial amount of energy corresponding to some hundreds of rads is sufficient, if applied to the whole body, to kill a human being.

It has become increasingly accepted by scientists in the last twenty years that, whatever the benefits we may obtain from radiation in the doctor's or dentist's office, all the radiation dose received by a person is ultimately harmful. Even the natural background radiation which all of mankind has received from the beginning of time may be responsible for some of the cancers which have occurred naturally since man first appeared on earth. Our knowledge of the effects of radiation on the body has been mainly derived from the case histories of people who have suffered malignancies as a result of exposure to radiation during their lives. These include miners who worked in the pitchblende mines at Joachimstal in Czechoslovakia and, after periods of twenty years or longer, developed lung cancer as a result of their exposure to radon gas and its daughter products in the mine. In the early days of radiology, when the hazards of radiation were not well known and virtually no safety precautions were observed, many of the radiologists suffered disfiguring injuries as a result of their exposure to X-rays. In the two decades following Becquerel's discovery of natural radioactivity in 1896, more than 100 research workers died of exposure to radiation.

In the 1930s many women, who had been employed during World War I painting luminous dials with paint containing radium, suffered fatal cancers of the jaw because of their practice of "pointing" the tip of the brush with their lips in order to achieve a neat result and thereby ingesting radium. In more recent times, much information has been obtained as a result of the study of the survivors of the Hiroshima and Nagasaki atomic bomb explosions, in whom approximately sixty cases of leukemia (additional to those which would have been expected in a Japanese population of similar age distribution who were not exposed to the radiation from the explosion) occurred.

It is apparent from all of this information that injury can be caused both by radiation from a source, such as an X-ray machine, located outside the body and also from radiation originating inside the body from radioactive materials which have been ingested by being breathed in air, drunk in water or eaten in food. The parts of the body mainly affected by the various radioactive isotopes are shown in Table 3.

The human body is rather a marvellous chemical factory which is even more advanced than the most modern existing factories for it is completely automated. Its loading docks are the mouth and nose and substances arriving there are instantly analysed and sent to the part of the "factory" where they are needed. Thus the body recognises calcium in milk and channels it to where it is required for building or repairing bone. However this sophisticated analytical department can be fooled. Barium, strontium and radium are elements which are chemically very similar to calcium and any ingested quantities of these will be treated by the body exactly as it treats calcium and will be laid down in the bone where they will remain.

Table 3

Parts of the Body Affected by Ingested Radioactive Isotopes

Iodine 131	Thyroid gland
Strontium 90, barium 140, lanthanum 140	Bones
Insoluble radioactive airborne emitting alpha, beta and gammas rays	Lungs
Ultra small, insoluble radioactive particles	Bone marrow, liver, and spleen
Caesium 137	Lymph glands and blood
Krypton 85*	Body fat
All soluble fission products (including some of the above)	Gonads and kidneys

*Krypton is one of the "rare" or "noble" gases, so called because they do not react with other elements. The main hazard from radioactive rare gases is from external radiation. Ingested rare gases are temporarily retained in body fat.

In the normal course of events this does not matter much. All bones contain a small amount of strontium and barium in addition to their calcium and this does not affect the bone at all. Radium is a different thing entirely, for it is radioactive, and radium laid down in the bone will continue to decay and irradiate one of the most radiation sensitive parts of the body — the bone marrow. Luckily people do not ingest much radium, despite the rather widespread occurrence of radium in soil, during their normal life. However two of the fission products formed in a reactor are bone seekers, strontium 90 (half life 28 years) and barium 140 (half life 13 days), and if ever they are released in quantity in an accident to a reactor, it is very important to ensure that they are not ingested by people.

The element iodine is also important, in trace quantities, to the normal operation of the body. It is used in only one place, the thyroid gland in the neck, where it promotes the formation of thyroid hormone, which regulates body growth and metabolism. It has been established that iodine 131 is one of the critical isotopes as far as ingestion of fission products is concerned.

Release of Fission Products from a Nuclear Reactor

Consider an accident to a reactor in which fission products are first released into the containment vessel and the containment is subsequently breached and the fission products are then injected into the atmosphere. There are two extreme cases: in one the radioactive materials are released at, or near, ground level and in the other they are injected into the atmosphere at some height above the ground. The first case causes greater problems close to the reactor and the second to a large extent mitigates the close-in effects at the cost of eventually affecting a greater area, although to a lesser degree. As the plume of fission products travels downwind it is first of all diluted by the wind itself. The cloud will also spread laterally and vertically, becoming more dilute the further it travels from the reactor. However the bigger it gets so the area affected by it becomes larger and larger (Figure 8).

What happens to the people in the affected zone? As the cloud travels overhead it delivers a radiation dose to the people below it. The size of the dose received will depend on the height of the plume (the higher the plume, the less the dose), the concentration of the radioactive materials in it, the isotopes present, and the time for which the plume persists (i.e., the duration of the release from the reactor). Of course the direction of the wind may change during the release and remove the radioactive material from above one area, only to cause it to pass over another. During the passage of the plume any large radioactive particles carried out of the containment will begin to fall out under gravity and they will reach the ground at a distance from the source depending on the size of the particle, its density, the wind speed and the height of the release.

As the plume travels farther from the reactor, smaller and smaller particles will reach the ground. Eventually it grows so large that it sweeps along the ground and, at this stage, deposition occurs, by impaction on surfaces (such as buildings, trees, crops and grass) and by turbulent diffusion to them, of those particles which are so small that they do not fall out appreciably under gravity, and of gases and vapours. Although gases and vapours are relatively unaffected by gravity they will nevertheless be deposited on both horizontal and vertical surfaces as the air sweeps them over the surface. A familiar example is the way in which the gas sulphur dioxide, emitted from coal-fired generating stations, is deposited on crops and buildings and causes damage. Molecular iodine vapour, for example, deposits out on any surface over which it is passing as if it were moving towards the surface with a speed of one or two centimetres per second.

As the plume sweeps along the ground people who happen to be there will be subjected to a radiation dose from the gas in which they are immersed, radioactive particles and vapours will deposit on their clothes and the exposed parts of their bodies, and they will inhale both particles and vapours. If the inhaled gases and vapours are still in their molecular state they will be deposited out rather quickly in the nose and pharynx, together with the larger of the inhaled particles. The smaller particles, together with any vapours absorbed on them, may be carried deep into the respiratory system and a considerable fraction of them may be retained for some time within the body and continue to give an internal radiation dose while they decay.

The internal hazard which is of greatest importance following a reactor accident is the ingestion of iodine 131 which may be inhaled (if the subject is immersed in the fission product plume) or ingested through the food chain. Iodine passing over the ground is deposited (or washed out by rain) on to grass; the grass is eaten by cows who concentrate it in their milk. The milk is drunk by people and the iodine concentrated in their thyroids where, in decaying, it gives a radiation dose to the thyroid gland. The thyroid is a small organ so it receives a greater dose than would be incurred if the iodine were spread throughout the body.

Babies and small children are especially at risk for two reasons, they have small thyroids so that the iodine is concentrated in a smaller space, and they drink a lot of milk. The radioactive half life of iodine 131 is eight days and its biological half life is five days (that is, half of any ingested iodine, radioactive or not, is eliminated by the body in five days). The short biological half life means that the damage done to the body is self-limiting, but the short radioactive half life means that the radiation from even a small quantity of iodine is intense. The danger is that people ingesting radioactive iodine may, as a result, eventually develop cancer of the thyroid. There is as yet no evidence of this having happened as a result of an accident to a nuclear reactor, but out of a group of 68 people on Rongelap Island exposed accidentally to fallout from an American hydrogen bomb test that went awry on Bikini Atoll in March 1954, 24 had developed thyroid nodules 22 years later and four of these turned out to be malignant. These people had received very large thyroid doses. It was estimated that the children in the group (because of their smaller thyroids) had received doses ranging from 700 to 1400 rem and the adults from 220 to 450 rem.

It is apparent from all of this that the most immediate objective of civil authorities after an accident to a nuclear reactor, which results in a release of fission products, must be to ensure that as few people as possible are exposed to direct radiation during the release and that people in the affected areas are prevented from ingesting contaminated food and liquids (particularly those containing radioactive iodine) after the release.

The second of these is very much easier to achieve than the first. It takes between two and three days for any radioactive iodine ingested by cows to work its way through the system and appear in milk, and this allows time for arrangements to be made to collect and dispose of contaminated milk. On the other hand the radiation dose from the plume starts to accumulate practically as soon as the release begins. Effective remedial action therefore depends on the prompt action of the reactor operators in informing the civil authority of the emergency, and secondly, on the existence of an emergency plan which can be put into effect very quickly. There is always a natural tendency on the part of the reactor operators to delay informing outside authorities of the accident, presumably in the hope that it may not be as serious as they think, or that their remedial measures may quickly get the situation under control, or even because they are conscious of the immense disruption caused to peoples' lives by being summarily evicted from their homes, maybe in the middle of the night. One can only advise them that it is better to be safe than sorry.

What To Do at Home

Let's assume that you live near a nuclear reactor and the unimaginable happens and you hear an announcement on your radio that there has been a release of radioactive material from the site. Your first reaction is probably to close all the doors and windows and this is a sensible thing to do. Then look out of a window and try to decide which way the wind is blowing. Unless it is blowing from the reactor towards you, you don't have much to worry about at the moment. Keep your radio on so that you will hear any further announcements. If the wind is blowing from the reactor towards your house, there is a possibility that you may be evacuated under the emergency plan and, to be ready for that, you should pack an overnight bag with any toilet articles and clothing you may need for two or three days. If any members of your family are at work or out of the house pack some things for them as well. If you should be evacuated before they return, they may not be allowed back to the house, but they will catch up with you.

Your house will probably be protected to some extent already against the entry of fission products. Since the energy crisis most people have sealed their houses quite effectively against loss of heat and entry of draughts. If you have not done so, you might be tempted to go around plugging up obvious leaks around door frames and window frames with a makeshift caulking of newspaper soaked in water, and also to close the flap on the chimney. If your house is effectively sealed and it is winter, make sure that enough air is available to allow the furnace, boiler or woodstove to operate properly. They need a lot of air and if the supply is restricted you run the danger of forming carbon monoxide, a deadly poison which is fatal within minutes at a concentration of one per cent in air. Many more people die of carbon monoxide poisoning every year than have ever been killed as a result of nuclear accidents. If your house is heated by a hot-air system change the furnace filter or switch on the electronic air cleaner if one is installed. Even if it is summer, switch on the circulating fan on your furnace (in conjunction with the filter or electronic air cleaner it will remove much of the fission products which are incorporated in or attached to particles). Finally, get a good book and sit down and read it, keeping the radio or television on, either on the ground floor in a place from which you cannot see the ground outside and on the side of the house away from the reactor, or, preferably, in the basement.

The reasons for this can be seen in Figure 9. If the plume of fission products has passed, or is passing over your house, the farther you are beneath it and the more solid material there is between you and the plume, the less radiation dose you will get. If the plume is already sweeping along the ground when it passes over your house, some radioactive fission products will be deposited on your roof. The reason for sitting on the side of your house remote from the reactor is that, as the plume passes over your house a large eddy is formed in it, as shown in Figure 9, and this results in an area immediately downwind of your house being practically free of deposition.

As soon as you hear on the radio that the emergency is over, or, as soon as you see that the wind has changed, so that it is no longer blowing from the reactor to your house, open all the doors and windows and keep them open for half an hour or so. **This is very important for it will ensure that any fission products which have entered your house are quickly flushed out of it.**

The advice in this section is based on measurements made by the author at Windscale after the accident in 1957 and on experiments subsequently conducted by him (Megaw, 1962). He discovered that when a cloud of fission products blows over and around a house, the amount that penetrates into the house depends on how well the house is sealed. This is fairly obvious so far! Every house, no matter how well it is sealed, has a certain ventilation rate which will change with wind speed and direction. The higher the wind speed, the greater the rate of ventilation. In a fairly well-sealed house the ventilation rate will

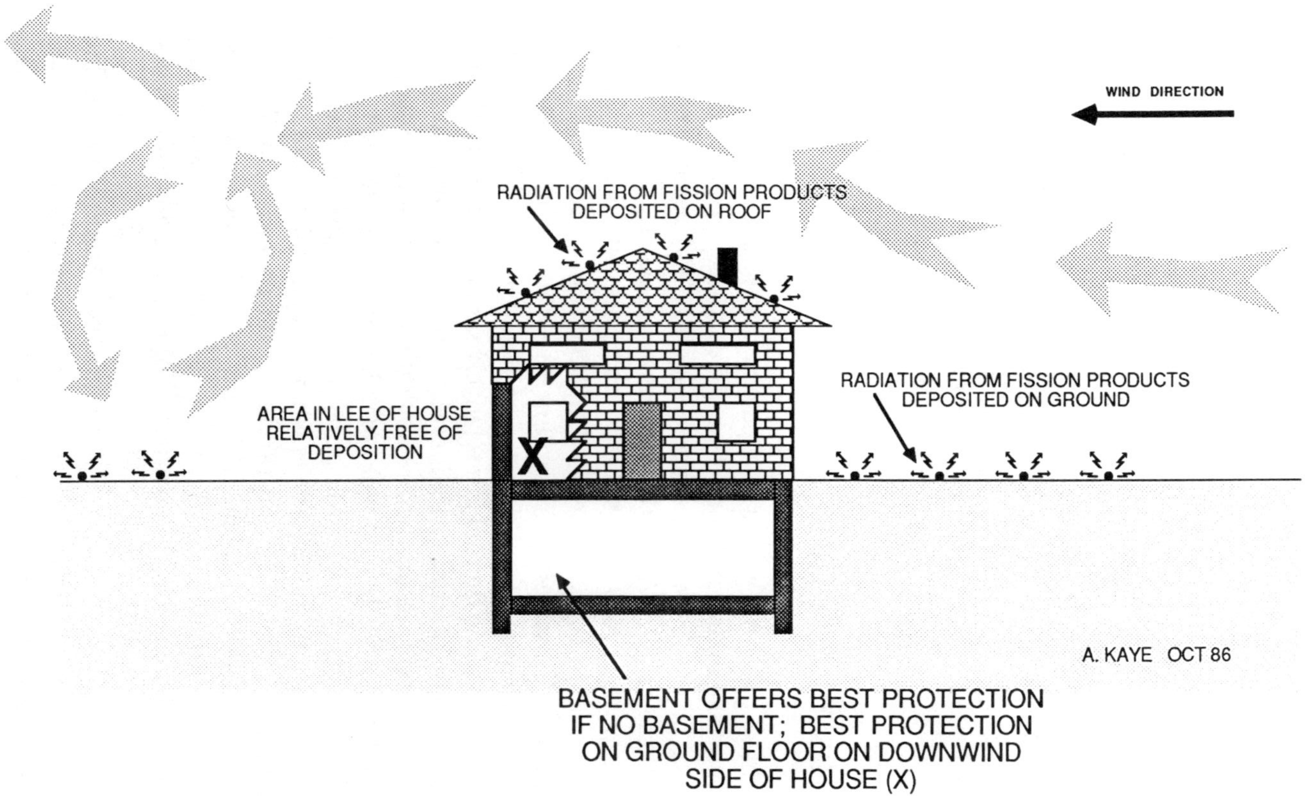

Figure 9 Idealised Behaviour of Fission Product Plume Passing Over a House

probably be between one air change per hour and one air change every ten hours. The concentration of fission products will, of course, be less in a house with only one air change every ten hours than in one with an air change every hour.

However the amount of fission products inhaled by an occupant doesn't depend just on the concentration of fission products in the air of the house, but rather on the concentration multiplied by the time for which it persists. In a house with a low ventilation rate, fission products will leak in slowly, but **they will also leak out slowly**, and, unless the doors and windows are opened promptly after the release is finished, the product of the concentration and time may be a high proportion in the low ventilation house of what it would be in the leaky house.

Potassium Iodide

You have probably read in the papers that it is a good thing to take potassium iodide if a nuclear accident occurs, and you may wonder, if you live near a reactor why no one has supplied you with these tablets. After the Chernobyl accident, according to press reports, it was impossible to buy potassium iodide tablets in Germany or Italy — they had all been sold to people anxious to protect themselves against iodine 131. The theory behind this is deceptively plausible. All iodine ingested by the body is routed to the thyroid; the thyroid is a relatively small organ and can only hold a certain amount of iodine; therefore if we can ensure that it is full of stable iodine before any radioactive iodine is ingested by the body, the thyroid will not take up any more iodine and any ingested by the body will be eliminated promptly. Therefore eating potassium iodide tablets when you hear that your nearby reactor has released fission products will protect your thyroid against uptake of iodine 131 and any taken in will pass harmlessly through the body. So the theory goes. In fact there is spirited debate over the advisability of taking potassium iodide pills.

In the United States the Food and Drug Administration and the American Thyroid Association have set standards for administering potassium iodide in a nuclear emergency. The FDA has recommended potassium iodide for populations expected to be exposed to sufficient iodine 131 to deliver radiation doses of 25 rem or more to their thyroids. The suggested dosages were 130 milligrams per day for those over one year of age and 65 milligrams for infants. In 1984 the American Thyroid Association recommended that the tablets be administered only if exposure delivered as much as 100 rem to the thyroid.

By comparison a medical X-ray used to be rated at 1.5 rem per six seconds of exposure although this should have been substantially reduced by more modern methods. However the New York Academy of Medicine has recommended against stockpiling potassium iodide for use in such an emergency in New York State. The recommendation was based on a number of factors. First was a lack of convincing evidence that iodine 131 causes thyroid cancer. Second were the adverse affects that potassium iodide would have on people with various thyroid disorders and those who may have an allergic reaction. These people constitute from two to three per cent of the population according to Dr. David Becker, chairman of the Academy of Medicine's committee on low-level radiation effects. It has also been suggested that pregnant women and small children may be harmed by eating potassium iodide tablets.

At a 1983 symposium on the problem Dr. Rosalyn Yalow, who shared the 1977 Nobel Prize in medicine and physiology for her work on the use of radioactive tracers (including iodine 131) in hormone research, endorsed the opinion of the Academy of Medicine. Discussing the potential problems associated with taking potassium iodide pills, she said that when a population is threatened by high levels of airborne radioactivity, "a more sensible plan would be to remain indoors, preferably in a building with minimum air exchange with the outside."

You are advised therefore **not** to take potassium iodide tablets unless specifically instructed to do so by the civil authority. It is especially dangerous to try to prepare oneself for a possible release of radioactive iodine by taking potassium iodide pills in advance of an incident, for each treatment is effective for only twelve hours. Above all, do **not** take any other forms of iodine. It was reported that, after the Chernobyl accident, many people severely injured themselves by drinking tincture of iodine, a corrosive poison, in the belief that it would protect them against iodine 131.

After the Emergency Is Over

When you have been informed that the release has ended or (if you have been evacuated) when you are allowed to return home, you may be concerned whether or not your house is safe or whether it is contaminated with radioactivity. Open all the doors and windows (if you have not done so already) and leave them open for half an hour or so to flush the house out with clean air. It is unlikely that surfaces within the house are seriously contaminated, for the experiments at Windscale, referred to earlier, showed that deposition of radioactive materials inside buildings was only a few per cent of that outside. However, to be sure, run over the house with the vacuum cleaner, take the cleaner outside and change the dust bag, seal the old one in a plastic bag and put it in the garbage. Wash your hands after this.

It is probable that, unless there has been heavy rain since the accident, there will still be fission products on the roof of your house and these may continue to give you a radiation dose, even though the release has long since ended. Wash your roof down with a garden hose, but do it from ground level and do not be tempted to go on to the roof. Apart from the danger of falling, going on the roof will bring you into close proximity to any radioactive materials on it and you may also contaminate yourself. If you have a hose attachment for spraying insecticide, put a little detergent in that and attach it to the hose. If there is any possibility of your house having been seriously contaminated your local authority will have arranged for it to be surveyed before allowing you to return to it. However, if you have any doubts at all about your safety, make a request for an additional radiation survey.

What to Do before an Accident

It may seem strange for this section to be at the end rather than the beginning, but it is not possible to plan in advance without having some idea of what may happen. The first, and most important thing is to ask your local authority for details of their emergency plan. If they haven't already got one it is time that they had. Experience has shown that the existence of a viable plan and its efficient implementation can do more than almost anything else to mitigate the effects of any disaster, whether natural or manmade. In the Mississauga, Ontario accident in 1979 in which a train drawing tankers loaded with potentially lethal chemicals was derailed and set on fire, more than a quarter of a million people were successfully evacuated from their homes without any panic ensuing or any injury being caused — an astonishing achievement.

If you live within 20 kilometres of a reactor it is worth knowing in exactly what direction the reactor lies from your house. It is also worth having a road map on a fairly large scale on which both your house and the reactor are marked. Buy a 30-degree set square (available from any store which sells students' supplies). Finally make a daily habit of being aware of the wind direction. If this implies a touch of paranoia, that is not the intention. In every major nuclear accident to date, Windscale, Three Mile Island and Chernobyl, the biggest failure of the authorities was in the timely release of adequate reliable information to the general public. It would be good to think that they will have learned their lesson by the time of the next accident, but probably better not to rely on it. With this simple set of aids you will be able, if ever an accident occurs to your nearby reactor, to determine pretty accurately the likely path of the radioactive plume.

Set the point of the 30-degree set square on the reactor and align the bisector of the 30-degree angle along the wind direction. The 30-degree angle will then cover the area likely to be affected by radiation and deposition from the plume. If your house lies outside this angle, you are unlikely to be affected unless there is a change of wind direction. If it is inside you should follow the course of action outlined above. Other precautions are fairly obvious. It is possible that water and electricity supplies may be interrupted even if your house is not in the area affected by the plume and also that milk supplies may be restricted. It is prudent therefore to have some modest store of candles, or other emergency lighting system, bottled water and dried milk. The most important thing to remember is to keep a cool head and not to worry too much. In spite of what all the nuclear power objectors say, the chance of a member of the general public being injured as a result of a reactor accident is extremely remote.

Chapter Four

The Radioactive World

Radioactivity is not a creation of modern man. The earth is radioactive and it always has been. All of our lives, from the cradle to the grave we are continually being irradiated, and so has every person who has ever lived on earth. The radiation comes from the sun, from outer space, from the ground we walk on, the houses we live in, the air we breathe and the sea we swim in. Before we can have any concept of the possible risk from either the normal operation of nuclear reactors or from accidents like Windscale, Three Mile Island or Chernobyl, we have to study the natural background radiation to which everyone is exposed.

Cosmic Rays

We are bombarded all the time by the high-energy-charged particles from the sun and from outer space which are known as cosmic rays. At sea level the average dose is about 30 mrem per year. By the time the cosmic radiation gets to the earth's surface, much of it has been absorbed by the atmosphere. It follows that the higher you go in the atmosphere the greater the dose of radiation you get from cosmic rays. Thus people who live in Colorado or New Mexico at five or six-thousand feet above sea level get a cosmic radiation dose about twice that at sea level. Passengers in aircraft also get enhanced radiation doses and at 12 km, the normal operating height for jet aircraft, the dose rate is about 0.5 mrem per hour, or about 150 times that at sea level. You might think that you can escape cosmic rays by staying indoors. To some extent you can, if you are prepared to accept an additional dose from the building materials of which your house is made.

Radioactive Substances in the Earth

All building materials derived from earth — bricks, concrete, stone, and, to a lesser extent, wood — contain radioactive substances, so that we also get a radiation dose from our houses. The earth itself when it was formed contained uranium 238, thorium 232 and uranium 235, all of which have half lives of the order of one to fifteen billion years, so that there are still plenty of them around. Each of them is the parent of a radioactive series which contains a dozen or so radioactive daughters. The decay chains of the uranium 238 and thorium 232 series are shown in Figure 10.

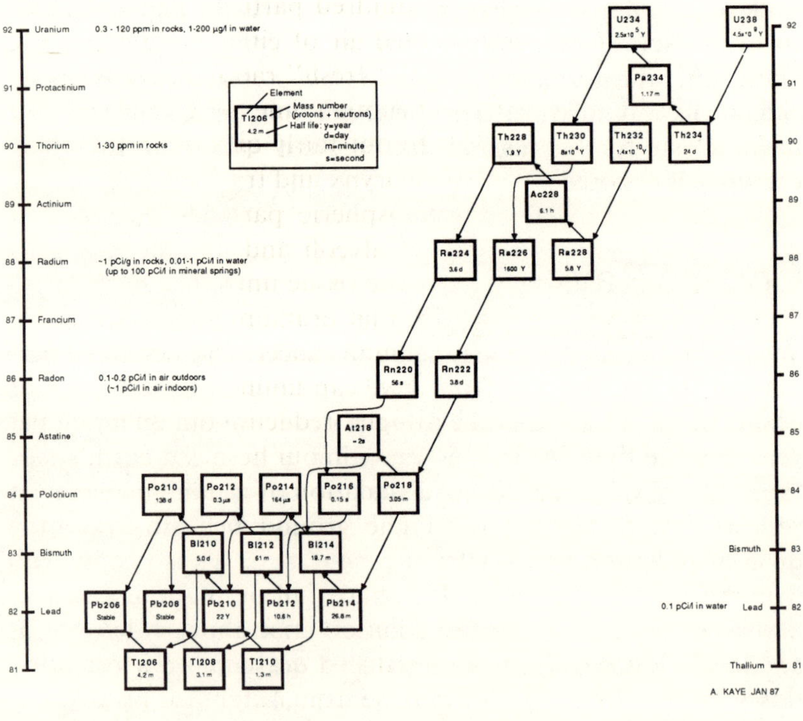

The uranium-238 and thorium-232 decay series

Figure 10 Decay of Naturally Occurring Radioactive Isotopes

Probably the most important of the daughter products is radium 226 (half life 1620 years) which decays into the radioactive noble gas radon (half life 3.8 days). Radon is emitted continuously into the atmosphere from the earth. It has daughter products which are all radioactive (with half lives ranging from a ten thousandth of a second to 22 years) and eventually decay to lead 206 which is stable. The air we breathe always contains radon and its decay products. Breathing radon in is bad enough, although its half life of 3.8 days means that very little of it decays while it is in our respiratory system.

The decay products are another matter. They are formed in the atmosphere as individual atoms but, after a while they become attached to the minute dust particles in the air. Even in an ultra-clean atmosphere in unpolluted Greenland or Antarctica there are usually several hundred particles per cubic centimetre of air and in the polluted air of cities there may be as many as a million. If we breathe "fresh" radon decay products (i.e., those which have not had enough time to become attached to particles), they are deposited out fairly quickly in the upper respiratory system (the nose, pharynx and trachea), but, if they have become attached to atmospheric particles they will be drawn deep into the lungs and alveoli and may be deposited there and continue to irradiate the tissue until they decay away completely. It is for this reason that uranium miners have until recently had a high incidence of lung cancer. The atmosphere in our houses contains radon and we can unintentionally increase that concentration by, for example, reducing draughts to conserve energy. On average we receive in our homes the equivalent of 80 mrem per year from the inhalation of radon and its decay products which diffuse from the ground beneath and from building materials.

In the days of the open fire almost all of this was removed rather quickly but with energy conservation the ventilation rate of the average home has been reduced drastically with a consequent increase in the radiation dose from radon and its daughter products. Someone living in a house with an electric heating system is likely to experience higher radon concentrations, and hence higher dose rates, than someone who lives in an oil- or gas-heated house where the need for combustion air provides a supply of ventilating air regardless of whether the owner wants it. The radiation dose received in different houses with different rates of ventilation has been studied in great detail in Sweden by Swedjemark (1978) and others quoted by him.

In wooden houses he found the average concentration of radon to be 0.015 Bq per litre and this did not change significantly even when the houses were not deliberately ventilated for 24 hours before the measurement (indicating that the natural ventilation rate of wooden houses is quite high). An unventilated brick house had a concentration of 0.05 Bq per litre and this dropped to 0.02 Bq per litre when the house was ventilated on the day of the measurement. Houses made of aerated concrete containing naturally radioactive alum shale had radon concentrations of 0.13 Bq per litre which was reduced to 0.06 Bq per litre by ventilation. In 63 houses of differing construction materials the concentrations ranged from 0.03 to 0.8 Bq per litre corresponding to a radiation dose to the occupants of up to 13 rem per year.

In spite of the cost of heating and the incentive to save money by preventing draughts, it makes sense not to seal up your house too tightly. The tightly sealed house is dangerous not only because of the possible buildup of carbon monoxide (page 75), but also because of the long-term danger from a high concentration of radon.

Radioactive People

As if this were not enough, our very bodies are naturally radioactive for they contain carbon and potassium and a proportion of these isotopes is made up of the radioactive isotopes carbon 14 (half life 5,500 years) and potassium 40 (half life 1.3 billion years), which occur naturally. The radiation dose from the potassium in our bodies is very much greater than that from the carbon 14. A person weighing 70 kg would contain about 140 g of natural potassium whose potassium 40 content gives a total dose rate of about 20 mrem per year.

Radiation from the Ground

The average radiation dose from the ground itself has been

estimated by the United Nations as 32 mrem per year and 95% of the world population receives doses of between 21 and 43 mrem per year. There are two rather exceptional areas. One is in Kerala in southwest India where the highly radioactive monazite sands give an average dose rate to the 70,000 inhabitants of 380 mrem per year with 6 per cent of these getting doses in excess of 1 rem per year. The other is in Brazil where average doses of 550 mrem per year have been reported in certain areas with a range of from 90 to 2800 mrem per year. The variation of this radiation dose rate around the world is shown in Table 4. It should be stressed that investigations of populations living in these high background areas have not shown any increased incidence of cancers. Such effects, if they exist, are so small as to be hidden by the normal variability of the occurrence of cancer, which may also be caused by environmental and genetic factors which are not related to radiation.

Table 4

World Variation in Natural Background Radiation Dose

	Annual Dose (in millirems)
US average	100
minimum (east coast)	90
maximum (Colorado)	250
UK average	110
minimum (London)	80
maximum (Aberdeen)	170
Sri Lanka granite areas	3000–7000
Kerala, India	800–8000
Minas Gerais, Brazil	1700–12000
Rio de Janeiro beaches, Brazil	550–1250
Cruising altitude of sub-sonic jet	3500*
Cruising altitude of supersonic jet	16000*

*resulting in a dose of about 4 millirem for a transatlantic flight.

Source: NRPB R77

It is clear therefore that everyone on earth experiences a radiation dose which is entirely natural and whose size will vary between quite large limits depending on where in the world he lives, what sort of a house he lives in, how well it is sealed and how it is heated.

In addition everyone receives an additional dose from what might be described as the "appurtenances of civilisation" — diagnostic X-rays, nuclear medicine, therapeutic radiation, radiation from television sets, luminous watches and clocks and from other devices incorporating radioactive materials such as fire detectors. Undoubtedly the largest of these is from the medical application of radiation.

Medical Radiology

The average dose received from diagnostic radiology per person will of course depend on the degree of technical advancement of the country concerned and also on the readiness of doctors in that country to send patients for X-rays. It is estimated that in the USA the mean bone-marrow dose from diagnostic radiology is about 100 mrem per person per year, in Canada about 60 mrem and in the UK about 35 mrem. In western industrialised countries the use of radiation in medicine has increased to the point where approximately half of the entire population receives X-rays each year for diagnostic purposes, and approximately half of all cancer patients receives radiation as a form of therapy at some stage during their disease. This adds up to a lot of radiation being given to a great many people, and should be a cause for considerable concern, although the very much smaller dose given by the operation of nuclear reactors seems to loom much larger in the media and in the public mind.

Biological Effects of Low Doses of Ionising Radiation

The effects of ionising radiation on the human body are threefold. First of all radiation can kill cells. This is why it is used to treat cancer. The aim of therapeutic radiology is to kill the cells in the tumour so that they cannot go on reproducing, without, at the same time, killing the healthy cells surrounding it.

Most popular concern about radiation is "the effects on future generations" and there is no doubt that radiation can produce mutations in the germ cells of both male and female animals and lower organisms, such as yeasts and that changes produced may show up in future generations. There is, however, no direct evidence of any genetic effect of radiation in humans.

The third effect of radiation is to induce changes in the cells of the body which may lead in the future to cancer or leukemia. The main problem in studying the effect of radiation on people is to determine the relationship between the radiation dose and the effect. One obvious starting point is to say that human beings have lived with the natural background radiation since time immemorial without any obvious effects and that there must therefore be some threshold dose below which radiation is harmless. However this is not necessarily true, for some of the cancers, leukemias and genetic defects which have occurred from the beginning of time may in fact be due to the natural background radiation. There is a popular "back to nature" movement which seems to assume that anything "natural" is bound to be good, but that is not always true.

Another approach is to say that all radiation is harmful and that the amount of harm is proportional to the dose received, that is, if the dose is doubled, so is the harm. A third possibility is that the dose and the effect are linked by a more complex relationship. It is really difficult to decide, precisely because the effects are so rare and it is not possible to do experiments on human beings, so that our knowledge must be based on experiments on animals and insects, (from which it is difficult to scale up the results to human beings), and on the subsequent life history of early workers in the field, whose dose history may not be known with any accuracy.

Other groups of people whose experience may be of value are those treated with radiation for various diseases such as ringworm of the scalp and ankylosing spondylitis. The great problem is that diseases such as cancer and leukemia usually take a number of years to develop whether the cause is cigarette smoking or a radiation dose, so that long-term studies are needed. The interpretation of these is by no means easy, and it is not difficult to arrive at a conclusion that may be wrong.

This is illustrated in Figure 11 which shows the results of a study at the Argonne National Laboratory of the incidence of cancer for the white population of each of the United States plotted against the natural background radiation dose. It appears fairly obvious from this figure that increasing the radiation background dose above 140 mrem per year actually reduces the chance of getting cancer, and that one way of reducing one's risk of getting cancer by 10 per cent is to add an extra 100 mrem per year to the background radiation level.

Perhaps this is indeed the case. No one knows. However an alternative explanation is that there are many causes of cancer, environmental, dietary, habitual (e.g., cigarette smoking) and genetic. The trend of results in the graph may also be explained by the fact that the states with high background radiation levels just happen to be rather sparsely populated, non-industrial areas where the air and possibly also the water are relatively unpolluted and where people perhaps have a less complicated and healthier lifestyle. What the Argonne results do show is that 30,000 more Americans are dying of cancer per year in the states with low background radiation than would have died had they lived in those with high background.

The Long-term Effects of Radiation

In the 1940s sufferers from ankylosing spondylitis were routinely treated in England with well-controlled rather large doses of X-radiation to the spine and subsequently showed an enhanced incidence of leukemia. (A control group of 400 patients who were not treated with radiation showed no increase, thus indicating that ankylosing spondylitis itself does not predispose a patient to leukemia). The doses received were in the range of 500 to 3000 rem to the spine. Of about 14,000 of these patients in Britain who were treated with X-rays, there were 25 deaths from leukemia additional to what one would have expected

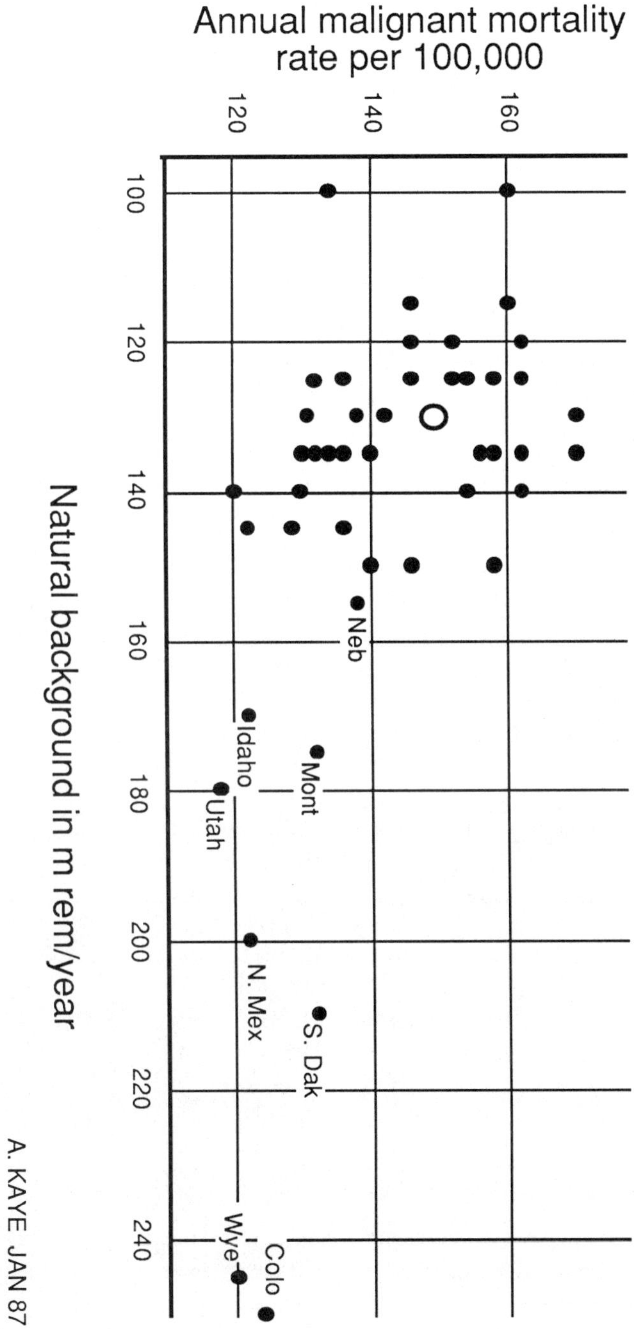

Figure 11 Malignant Mortality Rate in the United States as a Function of Natural Radiation Dose Rate.

Insert 5

Summary of Radiation-Linked Disease Surveys

Source	Radiation Type	Country	Number Studied	Tissue	Dose (in rads)	Excess Malignancies 1	2	3	4	5	6
Early radiologists	X-rays	USA	3700			11					
Miners:											
uranium	alpha	USA	3366	Lung	740				58		
uranium	alpha	Czech.		Lung	300				38		
non-uran.	alpha	Sweden	7500						20		
fluospar	alpha	NFLD	1100	Lung	200				61		
Luminous dial painters	alpha	USA	800					50			
Ankylosing spondylitis	X-rays	UK	14109	Bone Marrow	3210	25		3	37		
Thorotrast	alpha	various	4594	Bone Marrow	3500	54					
Ringworm	X-rays	USA	2872	Bone Marrow	30	5					
		Israel	10902	Thyroid	6.5		10	2			7
				Brain	120						
Thymus enlargement	X-rays	USA	2876	Thyroid	160	5	24				6
		USA	100	Thyroid	750		7				
			1452				56				
			859	Thyroid	725		57				
TB	X-rays		300	Breast	1000					20	
Fluoroscopy		USA	1047	Breast	150					18	
Mastitis	X-rays	USA	606	Breast	2100					7	
Ra 224 therapy	alpha	Germany	815	Bone	1000			47			
Benign disease therapy	X-rays	Sweden	1023	Breast	450					95	

Source	Radiation Type	Country	Number Studied	Tissue	Dose (in rads)	Excess Malignancies 1	2	3	4	5	6
Fetal examination	X-rays	UK	7649	Whole Body	30	10					
Hiroshima	gamma/ neutron		14657	Bone Marrow	10–200	50	23		26	12	
Nagasaki	gamma/ neutron		6740	Bone Marrow	10–200	13				6	
Marshall Islands	gamma/ beta		243	Bone Marrow	50	1	7				
				Thyroid	200						

1. Leukemia 2. Thyroid 3. Bone 4. Lung 5. Breast 6. Brain

(Source: UNSCEAR 1977)

Table 5

Risk Estimates for Radiation-induced Malignancies

Malignancy	**Risk Estimate** (number of cases per million people per rem)
Leukemia	15–25
Thyroid cancer (external radiation)	50–150
Breast cancer	50–200
Lung cancer	50
Bone tumours	2–5
Skin	dose > 1000 rad required
All other organs	low risk
All cancers	50–165 deaths

(Taken from Hall (1982) who quoted (i) Sources and effects of ionizing radiation, Report of the United Nations Scientific Committee on the Effect of Atomic Radiation, 1977 and (ii) the BEIR Committee Report: The effects on populations of exposure to low levels of ionizing radiations. National Academy of Sciences/National Research Council, Washington, D.C., 1972).

from the national leukemia death rate, an incidence of only 0.2 per cent, but even this low incidence was about ten times greater than would have been expected in the absence of radiation.

For the patients, one must weigh the benefit of relief from the pain of ankylosing spondylitis, a very painful and debilitating disease which affects young men, against the rather small risk of contracting leukemia in some years' time. Had the possibility of future ill effects been known at the time of irradiation it is probable that a considerable proportion of the patients would still have opted for the treatment.

A summary of the main classes of people in which radiation-related diseases have been observed is given in Insert 5.

Most of the data relating to production of cancer and leukemia in humans by radiation involve relatively small numbers of people exposed to rather large doses of radiation. In order to obtain risk estimates at low doses and to have some idea of the number of cancers produced when large numbers of people are exposed to very small doses of radiation, it is necessary to extrapolate from the high-dose to the low-dose region. The simplest assumption to make is that the risk per rem is the same at all doses. Based on this the risk estimates for radiation induced malignancies are summarised in Table 5. Using these, and bearing in mind that the values are probably overestimates, it is posssible to arrive at a rather crude estimate of the possible effects of any radiation-dose situation.

For example, the official Soviet estimate of the integrated radiation dose which will be received by the affected Russian population as a result of the Chernobyl accident is 29 million man rem. We have, from Table 4 that the estimated incidence of all cancers is between 50 and 165 deaths per million people per rem and so we can calculate that the number of additional cancer deaths in this population of a hundred million people over the next 70 years (assuming that the latent period for the cancers is 20 years), is between 29 × 50 (or 1450) and 29 × 165 (or 4785). During the same period it is expected that 9.5 million deaths will occur naturally from cancer in the same population.

Risk from Diagnostic Radiology

It is of some interest to apply these risk estimates to derive the risk from the radiation dose incurred by people in various countries from diagnostic X-rays. It has been estimated that the average annual medical dose rate in the USA is 100 mrem. As a result of this the number of cases of leukemia to be expected can be calculated by multiplying together the average bone-marrow dose of 100 mrem, the leukemia risk estimate of 15 to 25 per million per rem and the population of the United States, or $0.1 \times (15 - 25) \times 250$ or from 375 to 625 cases of leukemia per year. Based on the experience of the survivors of the Hiroshima and Nagasaki bomb detonations it might be expected that there would be three cases of cancer for each case of leukemia giving a total of perhaps 1200 to 1900 cases of cancer per year in the USA.

A report recently issued by the Harvard School of Public Health, Dartmouth Medical school and Boston Brigham and Women's Hospitals says that diagnostic X-rays are responsible for 250 cases of leukemia and 800 cases of breast cancer in women each year in the United States. These figures agree rather well with those given above for both sexes. A similar calculation for Canada gives 23 to 39 cases of leukemia per year and 70 to 120 total cancers. In the UK the comparative figures are leukemias 25 to 42 and total cancers 75 to 130.

It cannot be overstressed that these are very rough estimates and are probably very much on the high side, but it is not possible to be more accurate with our present knowledge. It will be argued that the benefit of a rapid and accurate diagnosis by X-ray far outweighs the rather small risk of contracting cancer. However there is a considerable spread between the average dose rates in the three countries and one wonders whether doctors in the UK are not using enough X-rays on their patients or whether those in the USA are using too many. Another disquieting fact is that the radiation dose received from the same procedure varies widely from hospital to hospital and this means that some patients are getting more radiation dose than they would if they had the same X-ray in a different hospital.

Genetic Effects

From the earliest days of radiation it has been a commonly held belief that exposure to radiation would cause untold congenital defects in generations to come. While insect and animal experiments show that radiation does affect and change genes it must be stated unequivocally at the outset that there is no evidence whatever of radiation having produced any genetic abnormalities in any human being, including the Japanese survivors of Hiroshima and Nagasaki, some of whom received very high doses indeed. All of our information on the genetic effects of radiation comes from experiments on the fruit fly drosophila or on mice.

Changes in genes are known as mutations. Mutations take place spontaneously in all creatures, including humans. No one knows whether they occur naturally or are caused by environmental effects such as the natural radiation background or chemicals present (whether naturally or artificially) in the environment. The extent of any genetic effect is measured by the effect it has on the normal mutation rate and, in particular, the radiation dose required to double the natural rate of mutation is termed the "doubling dose."

Again it must be stressed that there is no direct evidence in the human of mutations being produced by radiation. Such information as we do have is based on experiments with the mouse, especially in the "megamouse" experiment, so called because more than a million mice were used. The principal conclusions concerning the genetic effects of radiation are summarised by Hall (1982) as follows:

1. Only one to six per cent of spontaneous mutations in the human can be ascribed to background radiation.

2. Humans are not more sensitive and are probably less sensitive than mice, based on the lack of observable genetic effects in the descendants of survivors of Hiroshima and Nagasaki.

3. The male is much more sensitive than the female. Indeed, no significant increase in mutation rate was observed following irradiation of females at low dose rate, even for doses approaching the lethal level.

4. The number of mutations produced by a given dose of radiation decreases significantly if a time period is allowed to elapse between irradiation and conception.

5. A dose of 1 rem per generation increases the sponteneous mutation rate by about one per cent (60 to 1100 mutations per million liveborn compared to 107,000 occurring naturally.)

The genetically significant dose received by the general population due to diagnostic radiology has been estimated at 20 mrem per year in the USA and 11 millirem per year in the UK. The genetically significant dose is the dose to the reproductive organs received by people of child-bearing age and is thus less than the total average medical radiation dose. This is quite small compared to the dose received from natural sources which averages perhaps 100 mrem per year but still dwarfs the dose received from other man made sources e.g., radiotherapy 3, nuclear medicine 0.2, occupational exposure 0.8, fallout from weapons testing in 1954 – 1962 1.5, nuclear power (estimate for AD 2000) <1 mrem per year.

Summary of the Effects of Radiation on the Human Body

In summary it can be stated that it is probable that all radiation is bad for one, but that there is no evidence that the natural background radiation does any significant harm to human health or expectation of life. Any possible effect is so small that it is masked by other factors which have nothing to do with radiation. The natural background radiation dose is considerably higher in some parts of the world than in others without apparently affecting the health of the large numbers of people who live there. Nevertheless it is obviously prudent not to expose oneself to any more radiation than is necessary. The largest dose to which the average individual is exposed in the western industrialised countries is from diagnostic radiology. This undoubtedly results in cases of leukemia and cancers additional to those which would normally occur. The risk is rather a small one and in most cases is outweighed by the undoubted advantages of a rapid and accurate diagnosis. There would appear to be a considerable difference in the dose received in individual hospitals for the same radiological procedure and it appears only sensible to press for an improvement in techniques and equipment in those facilities which are at present giving dose rates greater than necessary.

Increased Risk at Low-dose Rates

In recent years several papers have been published which claim that there is a considerably greater risk at low dose rates than is admitted by the "radiological establishment." In every case in which these claims have been investigated there have been criticisms of the methods of survey and statistical evaluation used in the original work and the application of the original research results to other situations have shown that they do not reflect reality. Some of this work is discussed below.

Irradiation of the Fetus *in utero*

Stewart et al (1958, 1970) reported from a study of childhood cancer deaths around Oxford, that there was an increased history of *in utero* exposure to X-rays in children dying from malignant disease. According to these results there would be between 300 and 800 extra cancer deaths per million children per rem of exposure. In contrast a study by Court-Brown and Doll (1960) of 39,000 children irradiated *in utero* showed zero excess cancer deaths although between 11 and 31 would have been expected from Stewart et al's work. Jablon and Kato (1970) studied the histories of 1292 children born to survivors who were pregnant at the time of the atom bomb explosions. They found no excess of cancer up to age 10 whereas Stewart et al's work would predict between 5 and 14 cases for the 13.5 rem mean exposure.

The Mancuso Report

Mancuso et al (1977, 1978) analysed the relationship between 3500 deaths of cancer among workers at the Hanford Atomic Plant and their exposure to radiation as recorded on radiation film badges. They found an excess risk for cancer of the lung, pancreas, bone marrow and lymph system. Hutchinson et al (1979) reviewed this work and found there was no association with worker exposure for cancer of the lung, bone marrow or lymph system, or for all cancers. They did find a significant association for multiple myeloma and cancer of the pancreas. Their risk of cancer of the pancreas was 10 per million per year per rad which seems extremely high compared to other studies, for example 0.44 per million per year per rad for the ankylosing spondylitis cases and 0.83 per million per year per rad for the Nagasaki survivors.

Portsmouth Dockyard (New Hampshire)

Najarian and Colton (1978) studied cancer deaths in dockyard workers who were reported by next of kin as having been exposed or not exposed to radiation. They claimed an increased incidence of cancer in the workers exposed to radiation. The radiation records of the workers were subsequently released to the National Institute of Occupational Safety and Health which repeated the study and found that the number of cases was similiar to that found generally in employed populations of the same age distribution as the dockyard workers. The discrepancy is apparently due to misconceptions among the next of kin as to whether their relative was a radiation worker or not.

It seems fair to say that there is no reliable scientific evidence that low doses of radiation cause a health effect greater than would be expected from a linear extrapolation of the observed effects at higher doses. Nevertheless there is a great deal of disquiet in the public mind about the consequences of even minimal radiation, for example, from releases from nuclear power stations.

Alleged Effects of Releases from Three Mile Island

In spite of the fact that the releases of radioactive iodine in the Three Mile Island accident were very small there is still concern among people around Harrisburg that there were considerable effects on public health as a result of the accident. These were expressed by the then Mayor of Harrisburg, Mr. Stephen Reed when he gave evidence to the Public Enquiry into the siting of the Sizewell B PWR in England in November 1984.

He asserted that there were unexplained cases of thyroid "illness" downwind of Three Mile Island in Pennsylvania, that there was a doubling of the number of low-weight births due to the accident and that there was a 280 per cent increase in infantile mortality.

The Pennsylvania State Health Department investigated these allegations and reported that the "thyroid illness" cases were not related to the accident and in fact occurred in an area in the opposite direction to the prevailing wind at the time of the accident. There is no evidence of a doubling in the number of low-weight births due to the accident. Rather it has been found that there is a statistically significant correlation between low birth weights and excess medication. The 280 per cent increase in infantile mortality can only be derived by selectively drawing conclusions from a small sample.

The more comprehensive statistics published by the Pennsylvania State Health Department actually show a fluctuation in the other direction. The Department has found no evidence of any Three-Mile-Island-related effect on the rate of infant mortality. Nevertheless there appears to be a widespread local belief that the accident had a deleterious effect on the health of nearby residents. A later and even more bizarre suggestion is that, while the radiation from the airborne plume may not have caused any harm, nevertheless the plume "touched down" on nearby hilltops and caused health damage there. This is also unsubstantiated by any reliable statistics.

A couple who lived near Three Mile Island, Norman and Marjorie Aamodt conducted a survey which claimed to have shown a dramatic rise in cancer deaths clearly tied to the Three

Mile Island accident. An investigation of the allegations by the Department of Health found that cancer deaths which the Aamodts attributed to Three Mile Island included people who had been diagnosed as cancer patients before the accident, long-term heavy smokers who had died of lung cancer, persons who moved into the area after the accident and others who were mistakenly counted as residents of the survey area. Included in the cancer deaths was one person who had died of causes other than cancer and had apparently been confused with a relative who had died of cancer prior to the accident.

The Department identified 144 cancer deaths since the accident in the downwind communities of Goldsboro, York Haven, Fairview and Newberry which includes the neighbourhoods canvassed by the Aamodts. The figure is almost exactly the same (142) as professional epidemioligsts said would have been expected had the accident not occurred. For the area within a 10-mile radius of Three Mile Island the Department found that 2892 deaths had occurred from cancer between 1979 and 1985 which is slightly less than the 2909 expected by the epidemiologists. In a 20-mile radius 7924 deaths from cancer occurred while 8177 would have been expected.

Leukemia in Seascale

Seascale is a small village on the West Cumbria (formerly Cumberland) coast of England. It is on the edge of the Lake District and its population in 1945 was 630. It was a rather select seaside resort with a hotel, the Scafell, and a few shops. It later acquired some notoriety when the first nuclear reactors in Britain, the Windscale reactors, were built at Sellafield, three miles north of Seascale, and most of the scientists and managers at the plant lived in Seascale. The author and his young family lived there from 1951 to 1954. On 1 November 1983 Yorkshire Television transmitted a programme entitled "Windscale — the Nuclear Laundry" in which it was claimed that there was an increased incidence of cancer in the neighbourhood of the Sellafield site. As a result the British Minister for Health invited the past president of the Royal College of Physicians, Sir Douglas Black, to head an independent enquiry "to examine the evidence concerning the alleged cluster of cancer cases in the village of Seascale, to consider the need for further research and to make recommendations."

In its report the Advisory Group tried to reach a conclusion on the suggestions that an increased level of radioactivity in the neighbourhood of Sellafield has caused an above average incidence of leukemia. They say it is possible, but by no means proved. Their report considers the evidence about radiation doses to young people in Seascale over the years and concludes that the calculated dosages do not support the view that radiation from Sellafield was responsible for the observed incidence of leukemia there.

The report stressed however that there were unavoidable uncertainties in estimating the doses actually received by the young people and concludes that the proposition that the leukemias were caused by radiation cannot be completely discounted. One of the recommendations of the Group was that a designated body with significant health representation should be set up "to enable decisions on action with regard to the control of permitted radioactive discharges to take account of all relevant factors." As a result of this recommendation the Committee on Medical Aspects of Radiation in the Environment (COMARE) was set up. This committee produced its first report in August 1986. Additional information available to COMARE included the fact that the original estimate of a total of 400 grams of irradiated uranium released through the original faulty Windscale filters had been revised upwards to 20 kilograms. Hawkes et al (1986) suggest that this accounts for the discrepancy between the maximum of 0.091 excess leukemia deaths arrived at by the Black Group and the 3.5 excess deaths which occurred.

This is not so. The COMARE report points out that the Black calculation assumes that all leukemia cases in young people under the age of 20 in England and Wales are due to the natural background radiation, and this almost certainly produces an overestimate. It also shows that the increased doses received during the thirty-year period (from 1950 to 1980) by residents of Seascale are still well below the doses that are estimated to have been received by the population during this period from natural background and nuclear weapons testing fallout. Three separate independent analyses of the data available which approached the dose and risk estimation from different points of view and gave emphasis to different aspects of the data arrived at dose and risk estimates to the children born and brought up in Seascale which were remarkably similar and around 0.1 additional cases of leukemia. It is possible, although very unlikely that these are all wrong. It should also be pointed out that there are other areas in England, for example Slough in Berkshire and Dane End in Essex which also have "pockets" of excess leukemia incidence but have no apparent connection with nuclear plants or with local sources of ionising radiation.

As has been seen, there have been many claims that very low doses of radiation give rise to observable cases of disease. There will undoubtedly be many more. Each claim must be carefully examined by the bodies responsible for radiological protection — the International Committee on Radiological Protection, the BEIR (Biological Effects of Ionising Radiation) Committee of the US National Academy of Sciences and the Scientific Committee on the Effects of Atomic Radiation of the United Nations (UNSCEAR). Each of these organisations, made up of scientists experienced in the field of radiation protection, has come to the conclusion that the best available estimate of the risk of cancer from low doses of ionising radiation is that given in Table 5 and derived from the linear extrapolation of observed effects of high doses.

Is it likely, or even possible that they are all wrong? The evidence is that, if they are wrong, they have probably erred on the side of conservatism, that is, they have over-estimated the risks. Nevertheless, no one is complacent and further studies are continually being done. Even in populations subjected to high doses of radiation it is only the rare individual who suffers ill effects and at very much lower doses it will always be extremely difficult to detect any relationship between exposure and effect, particularly in the presence of other carcinogenic agents. There is no doubt that high doses of radiation are harmful but there is considerable doubt as to whether small doses have the effect predicted by linear extrapolation, or no effect or an effect somewhere between the two extremes.

Chapter Five

The Effect of Chernobyl on Other Countries

Many countries experienced increases in their level of radioactivity as a result of fallout and washout from Chernobyl. The level of deposited radioactivity was greatly enhanced when rain fell during the passage of the fission-product cloud. It is a common reaction to assume intuitively that just because the wind at the start of the accident carried the first detected radioactive material to Sweden, all subsequent travel of the emissions would be in that direction. This would be true only if the wind direction had remained constant for the duration of the release and until all airborne radioactive material had been deposited. In fact, every change in wind direction after the start of the accident not only changed the direction of travel of the fission products emitted from that time onwards, but also spread out all the fission products airborne at that time over a new, and greater, area. This is apparent from Figure 5 which shows the spread of fission products from Chernobyl.

The major releases from Chernobyl were spread over a period of eleven days and because of changes in weather patterns during that period, many countries, Finland, Sweden, Denmark, Germany, France, the United Kingdom, Italy, Austria, Poland and Czechoslovakia and, to a much smaller extent, Japan, the USA and Canada, were affected to a greater or lesser degree.

Extensive analyses of the weather situation over the relevant areas of Europe during the Chernobyl release were carried out by many agencies, in particular by the Finnish Meteorological Institute (STUK, 1986) which calculated the trajectories of air masses originating over Chernobyl at various times after the start of the release. It estimated that an air mass which was over Chernobyl at the start of the release would have reached the south coast of Finland by noon on Sunday 27 April and the northern areas of mid-Finland by late afternoon. There they met a northeasterly Arctic air mass which effectively blocked further penetration northwards.

During Monday and Tuesday morning the emission plume was still travelling from Chernobyl towards Finland. These fission products had left Chernobyl before noon on Sunday. By Sunday afternoon (36 hours after the start of the accident) the winds at Chernobyl were from the east so that the fission products were by then travelling to the west towards Poland and Czechoslovakia and this continued until 3 pm on Monday 28 April when westerly winds began to carry the fission products to the east, further into the Ukraine.

By the early hours of Tuesday 28 April a northeasterly wind was carrying the emissions into Rumania but the wind changed again by early afternoon and the radioactivity headed back east again. Releases during the morning and afternoon of 30 April were carried towards the southwest and southeast respectively and during the whole of Thursday 1 May a westerly wind carried the contamination deep into Russia and this pattern prevailed during 2 and 3 May.

The variability of the wind direction during the first week after the start of the accident thus resulted in a very much larger area of western and eastern Europe being contaminated, but to a very much lesser degree than would have been the case had the wind blown in a constant direction during the whole of the release. It is a moot point whether this is a good or a bad thing as far as agricultural damage or public health is concerned as it depends largely on the distribution of agricultural land and the population density in the directional sectors around the source.

A summary of the effects, obtained, where possible, from the official agencies of the countries concerned, is given below.

Finland

The first indication of the coming fallout was detected at the radiation monitoring station of Kajaani, in central Finland on Sunday evening, 27 April. There had been a shower of rain that afternoon. It was at first thought that the rise in radiation dose to 0.07 to 0.1 mrad/hour could have been due to something which had been observed in previous years, the spring release of radon on the melting of the winter snow cover. By Monday afternoon news of the measurement of fission products at Forsmark reached Finland and it was realised that that there was an extensive fallout situation. By Tuesday 29 April a considerable rise in the external radiation level was detected, first in the western parts of the country and later in southern and central Finland. The highest sustained external dose rate was 0.4 mrem per hour (about 40 times the normal background level) at Uusikaupunki, in south west Finland, on Wednesday 30 April. After this the levels decreased with a calculated half period of about six days. Measurements of the relative quantities of the various isotopes showed that the most abundant was iodine 131 (relative concentration 100) followed by iodine 133 (27), tellurium 132 (17.1), caesium 137 (5.5), caesium 134 (3.2), and barium 140 (2.6).

During the period following the release, a number of recommendations was issued by the Finnish Centre for Radiation Protection and other authorities. It was recommended, for example that rain water not be used for drinking or given to stock (2 to 15 May), that the sowing of fast-growing vegetables such as lettuce and spinach should be postponed (7 to 15 May), that cows not be allowed out to pasture (7 to 26 May), that children not be allowed to play in puddles (7 to 16 May). On 12 May it was recommended that air filters in houses be replaced. No restrictions were placed on the use of water from wells or waterworks or on staying outdoors. It was not thought necessary for pregnant women to take any special precautions.

The Finnish Centre for Radiation and Nuclear Safety has estimated that the total average additional dose to the Finnish population as a result of the Chernobyl accident will be 0.42 rems. This includes all dose pathways. This is about the same as an additional year's background radiation. Applying the factors given in chapter four suggests that a total of 38 to 63 additional cases of leukemia, and possibly three times as many cases of cancer may eventually occur in Finland as a result of the accident.

Sweden

The preliminary report of the Swedish National Institute of Radiation Protection states that the fallout from Chernobyl was first detected on Monday 28 April when personnel from the morning shift at Forsmark measured increased amounts of radioactivity on workers passing the entrance monitor at the station. The reactor site was partially evacuated, but it soon became clear that increased levels of radioactivity could be measured all over the east coast of Sweden. Measurements by the National Defence Research Institute showed that the source of the release was not a nuclear bomb but rather a release from a nuclear reactor. Backward air trajectories are automatically calculated by the Swedish Institute of Meteorology and Hydrology and they indicated that the radioactive materials were coming from the direction of Lithuania, White Russia and the Ukraine. Suspicion at first fell on the nuclear power plant at Ignalina in Lithuania, which is much closer to Sweden than is Chernobyl.

Extensive measurements of background radioactivity are made routinely in Sweden and these were intensified in the days following the accident. The highest sustained radiation dose was about 0.4 mrem per hour at the city of Gävle, about 160 km north of Stockholm, although in isolated smaller areas levels of up to 1 mrem per hour were detected. The area which had the highest contamination experienced rainfall on Monday and Tuesday 28 and 29 April. The Swedish National Institute of Radiation Protection recommended that there was no reason to take iodide tablets "since the negative secondary effects in the form of allergic reactions could be expected to lead to greater problems which would not be offset by a reduction in radiation doses." There was no need to remain indoors, pregnant women had no cause for anxiety. People were warned not to drink rain-water, but no restrictions were placed on drinking water from normal water supplies or from private wells.

It was not found necessary to restrict the distribution of milk, although it was recommended that cows intended for milk production should be kept indoors until measurements on grass were completed (at this time of the year most cows in Sweden are kept indoors anyway). It was recommended that vegetables grown in the open should not be eaten for a few days. This included nettles and dandelions "except that boiled nettles can be eaten if the first water in which the nettles are boiled is discarded" (a similar recommendation was made in Finland). Imports of meat, fish, potatoes and vegetables from the Soviet Union were stopped. Imports from other East European countries were allowed subject to checking samples for radioactive contamination.

The Swedish National Institute for Radiological Protection (NIRP) estimated that people in the most affected part of Sweden would be exposed to an additional dose of about 4 rem over a 50-year period as a result of the Chernobyl accident. That exposure would increase the cancer risk of those people by about 1%, but the NIRP report does not give any indication of the number of people likely to be exposed to that radiation dose. It does however give the present risk of incidence of cancer, ranging from 54 per 10,000 males (39 for females) in the county of Malmohus to 33 per 10,000 for both males and females in the county of Jamtland and estimates that these would rise by 0.3 per 10,000 as a result of Chernobyl fallout. Judging by the distribution of radioactive contamination (Figure 6), Sweden was at the western edge of the initial plume and the number of people receiving the maximum dose must be quite small.

Although the deposition of radioactive material was less in the northern parts of Sweden than around Stockholm, caesium deposited on lichen and eaten by reindeer has already had a serious impact on the lives of people in Swedish Lapland. The herding of reindeer is a traditional way of life for the Laplanders and, by August it had already been found that reindeer meat was so seriously contaminated with caesium 137 as to be unfit for human consumption. The acceptable level of caesium 137 in reindeer meat in Sweden is 300 becquerels per kilogram and it has been reported that many reindeer contain up to 15,000 becquerels per kilogram. It was estimated that 40,000 reindeer would have to be killed and buried and Lapland authorities consider that as many as 100,000 reindeer may have to be disposed of in the next five years. The Swedish government has undertaken to pay compensation to Lapps for financial loss caused by radiation from Chernobyl but amounts are still subject to negotiation. It has also been reported that Swedish farmers are claiming compensation for having to keep cows indoors and for the contamination of pastures.

Denmark

The initial impact of the Chernobyl accident on Denmark seems to have been rather less than on Sweden and Finland and suggests that only the western edge of the first plume passed over the country. Reports issued by the Danish National Agency of Environmental Protection show that the first radioactive materials from Chernobyl arrived over Denmark at noon on Sunday 27 April. As a result the gamma radiation background at the Riso National Laboratory rose from its normal level of 7.2 to 8.1 microrontgen per hour. This level gradually decreased to nearly its normal level during the next few days until about midday on Wednesday 7 May, when, during a period of rainfall, the gamma dose rate rose sharply to more than 10 microrontgen per hour. This sudden rise is obviously due to the second major release from Chernobyl which peaked on 5 May. The gamma dose rate at Riso then began to decrease gradually, although on Saturday 10 May it was still nearly 35% above its normal level.

The National Agency for Environmental Protection estimates that the total extra dose experienced by the Danish population as a result of the Chernobyl accident, including the future dose, from the radioactive material deposited up to 28 May "corresponds at most to approximately one month of natural background radiation." An interesting paper by J. Roed of the Riso National laboratory gives estimates of the velocity of deposition of iodine 131 and caesium 137 on plastered walls and concrete and asphalt road surfaces. The values of 0.02 cm/sec reported for iodine 131 suggest that, as expected, by the time it reached Denmark, the iodine was in some form other than elemental iodine vapour (for which the velocity of deposition is 1 to 2 cm/sec, depending on the conditions). At Windscale, also, the velocity of deposition of iodine 131 decreased with increasing distance from the reactor as the iodine progressively changed its chemical form (due to reactions with other chemicals in the atmosphere) or became attached to small particles.

The deposition velocities measured for caesium were about a factor ten less than those for iodine. Roed also estimated the protection given by remaining indoors and found that the inhalation dose incurred by someone who remained indoors with windows and doors shut was reduced by a factor 2.6 for iodine attached to particles or for elementary iodine vapour, but not reduced at all for methyl iodide, (a fraction of elemental iodine vapour released to the atmosphere changes to methyl iodide, CH_3I). Even more interesting, Roed measured the relative deposition in city, suburban, rural and wooded areas. He found that the relative deposition of iodine 131 was as follows: city 0.09; suburbs 0.58; rural areas 0.53; wooded areas 0.40. One reason for this is probably the well-known urban "heat island" effect which would tend to cause the plume to rise as it approached the city.

Italy

A very comprehensive account of the radiological consequences in Italy of the Chernobyl accident has been given by the Italian Atomic Energy Authority (ENEA DISP (86) 1). This shows that the first radioactive materials from Chernobyl arrived in the northeastern regions of Italy on 30 April. In the following days with changing wind directions, additional quantities of radioactivity arrived over the whole of Italy. The maximum levels were reached in Northern Italy on 1 and 2 May and, at lower levels, one or two days later in central and southern Italy. Maximum concentrations in plants occurred some days later than the maximum concentrations in air. Warnings of the need for careful washing of fruit and vegetables and to feed animals on dry forage were given on 1 May and advice to avoid consumption of fresh vegetables was issued on 2 May.

The external gamma dose rate in Italy showed little increase above background except in areas where heavy rain led to greater washout of fission products. Efforts were therefore concentrated on measuring the levels of iodine 131 and caesium 137 in air, vegetables and cow's milk. Iodine concentrations were between ten and twenty times those of caesium during the first week of May but, as time went on, they became much closer together. The other radioisotopes were almost always present at levels below that of caesium; for example the concentrations of strontium 89 and 90 were between 10 and 100 times less than those of caeisum.

The spatial distribution throughout Italy of the estimated average doses to the population from ingestion of these radioisotopes was calculated on the basis of data assembled up to 25 May. The maximum doses were incurred in northern Italy with lower doses in central and southern regions. The total future dose commitment consequent on the Chernobyl accident was estimated as about 5 mrem for adults and the effective total dose commitment for the Italian population as a whole as about 500,000 man rem. The report concludes "that it does not seem justifiable to proceed from this figure to an estimate of the expected health effects, which it does not seem possible to express in numerical terms." An estimate was also made of the total radiation dose received by a person spending 8 to 10 hours per day out of doors and taking into account the dose from direct irradiation from the ground, the dose to the skin by beta radiation, the dose from ingestion of contamination collected on the hands and the dose from contamination resuspended in the air. The effective dose turned out to be about 0.1 millirem per day or about 25 per cent of the natural background dose.

The United Kingdom

Radioactivity from Chernobyl was first detected in the United Kingdom on 2 May when airborne levels in the south of England rose to a maximum during the afternoon but declined to near normal levels the following day. Similarly increased levels were detected in Scotland on 3 and 4 May. Deposition in the south of England occurred in mainly dry conditions but in some areas of the north of England, Wales and Scotland there was sporadic heavy rainfall and this led to quite high concentrations of radioactivity in rainwater. The highest level recorded was about 30,000 Bq of iodine 131 per litre of water in a puddle in the north of Scotland, but the levels in rain collected in scientific instruments were generally much lower. The highest levels were above the derived limits calculated by the National Radiological Protection Board for drinking water and members of the public were advised not to drink rainwater. At all times the levels in piped and well water were well below that at which any restrictive action would have been necessary.

Levels of iodine 131 in milk also varied with location, primarily due to differing rainfall in different areas. In the south of England the concentration of iodine 131 in milk was about 50 Bq per litre while in the north a representative value was about 400 Bq per litre. Both of these values were well below the level at which the NRPB recommends that a ban on the sale and consumption of milk should be imposed. On 20 June the British government placed a three week ban on the movement and slaughter of sheep in northern England and Wales, because of high levels of caesium 137 in the meat. These high levels were obviously due to the enhanced washout of caesium during heavy rain in the hill districts of England and Wales where sheep farming is extensive, and its subsequent ingestion by sheep.

It was estimated that the total average additional exposure in the United Kingdom as a result of the Chernobyl accident was about 7 millirem which adds about four per cent to the average 200 millirem per year received from natural and man-made sources.

The total collective dose which will be received over the next fifty years by the population of the UK has been calculated as about one million man rems or an average dose of 0.02 rems per person. This may result in an additional 50 cases of leukemia and perhaps 125 cases of cancer in the next 50 years. By the same calculation the natural background radiation would be responsible for 60,000 cancer deaths out of the total expected in that period of seven million.

France

The radioactive plume from Chernobyl was first detected over Paris on 29 April. Twenty short- and long-lived isotopes were detected, the most prevalent being tellurium 132, followed by iodine 131 and caesium 137. After three days rain washed out more than 99 per cent of the radioactive materials but there was a subsequent increase (between 4 and 5 May), although not nearly to the original levels. A further increase between 10 and 19 May is ascribed to a second passage over France of the radioactive plume (with its radioactivity reduced by a thousand times) after passing around the earth. It was estimated that total deposition of caesium 137 in the Paris area as a result of the Chernobyl accident was comparable to that which occurred from fallout from nuclear weapons testing in 1963.

Japan

The first indication of the arrival of the Chernobyl plume in Japan was the abrupt increase of the radioactivity of surface air and water on 3 May, a week after the start of the Chernobyl accident. The concentration of iodine 131 was about one fiftieth of that observed in Sweden. The airborne activity decreased with an apparent half life of three days until it was reduced abruptly after a rain storm. Concentrations of radioactivity in rain followed those observed in the atmosphere and the highest concentration in rain (163 Bq per litre) occurred on 5 to 6 May and decreased steadily thereafter. The highest deposition of iodine 131 occurred on 6 to 7 May, presumably because of the relatively high rainfall on that day. The concentrations of iodine 131 in air were similar to those caused by a high-yield nuclear test by the Soviet Union in October 1961 and the total deposition of caesium 137 was about 2 per cent of the cumulative deposition caused by nuclear testing since 1959.

Eastern Europe

Little information on radiation levels in eastern European countries was reported in the media in the days following the Chernobyl accident. In a news conference in Warsaw on 30 April, Zbigniew Jaworoski, Director of the Radiation Control Institute said that radiation measurements were declining, after having reached dangerous levels. He said that the highest levels of radioactivity had been recorded at a village in the Masurian Lakes district, where the dose was 2.5 millirem per hour, or about 500 times the natural level. Mieczyslaw Sowinski, head of Poland's Nuclear Agency, predicted that there would be an increase "at a level of a few per cent" of thyroid cancer cases within the next 30 years. Dr. Krysztyna Bozkowa, Director of the Mother and Child Institute, said that the government might have overreacted by ordering iodine to be issued to children. She said that the decision was made at a time when no one knew how long the emissions of fission products from the reactor would continue.

Probably the most extensive and reliable measurements of the radiation levels in Eastern Europe are due to the UK Foreign and Commonwealth Office which asked the NRPB on 28 April for an appraisal of the radiation situation on the basis of which advice would be given to embassy staff and residents in affected countries. With what can only be described as a flash of brilliance the NRPB provided embassies in the affected countries with simple but sensitive dose-rate meters which were calibrated before dispatch. The embassies at Moscow, Warsaw, Bucharest, Budapest, Prague, Sofia and Belgrade were sent the instruments and also asked to provide samples of tap water, fresh milk, vegetables and grass for analysis.

The most significant radioisotopes in the samples of food were, as expected, iodine 131 and caesium 137. There was a wide variation in the concentration of iodine in milk, probably due to the pattern of rainfall, and the highest levels reported, from Poland and Hungary, were about 2000 Bq per litre which is the level at which the sale of milk would be prohibited in Britain. There was also considerable variation in the concentration of iodine and caesium in vegetables, but, interestingly, it was shown that concentrations in leafy green vegetables could be substantially reduced by the removal of outer leaves, washing and general kitchen preparation. The total dose received in a year due to the accident was estimated (assuming that people spend 90 per cent of their time indoors and that shielding by buildings reduces exposure by a factor 10) as being between 10 millirems in Warsaw and Hungary and a hundred millirems in northeast Poland and Rumania.

The USA

The levels of radioactivity found in air, rainwater and milk were generally much lower in North America than in Europe. Fission-product radioactivity from the Chernobyl accident was first detected at Oak Ridge, Tennessee on 10 May. Two distinct phases were noticed, the first lasting from 10 to 17 May and the second, in which the ratios of fission products were different, starting on 18 May. It is apparent that the two phases are associated with the early release (26 April to 1 May) and the later release (1 to 5 May). Traces of iodine 131 and caesium 137 were detected in milk from northwest Oregon on 11 May.

Canada

Iodine 131 was detected in rainwater which fell in Ottawa during a 24-hour period ending at 10 am on 7 May. The concentration was six times the acceptable level for drinking water and people were warned not to drink rainwater. Traces of iodine and caesium were also found in air sampled during the same period. Samples of rainwater taken in eight other towns and cities from Vancouver to Halifax showed no radioactivity. One of Greater Vancouver's three main reservoirs was closed for a few days due to a high reading of radioactive contamination which was later shown to be erroneous.

Low levels of iodine 131 (less than 1Bq per litre) were found in milk samples in Toronto, Ottawa and Fredericton, six times lower than the levels detected in 1964 as a result of fallout from nuclear weapons tests. Monitoring of Canadian grown vegetables showed no sign of contamination but customs officers in Toronto seized a shipment of French lettuce and herbs which exceeded the maximum acceptable level of 70 Bq per kilogram.

The effective radiation dose equivalent incurred by Canadians in the year following the Chernobyl accident has been estimated by Huda et al (1986) as 28 microrem. This is about ten thousand times less than the average Canadian will incur from the natural background radiation in the same period. Using the same assumptions as before (that the dose-effect relationship can be extrapolated linearly downwards from high dose rates) it can be estimated that the effect of the Chernobyl accident on Canadians may be 0.1 serious effects (cancers and genetic effects through two generations) occurring in the total population of Canada. The total radiological harm inflicted in Canada will therefore be negligible. The total deposition of iodine 131 on Canada has been estimated by Tracy (1986) as 26,000 curies and that of caesium 137 and 134 as 11,000 curies. There will be no agricultural impact on Canada.

The Aftermath

In an important paper John Garland and Roger Cambray of Harwell report that after about 20 May monitoring stations from all over the Northern Hemisphere (including places as far apart as Tromso in Norway, Gibraltar and Hong Kong) were reporting air concentrations of caesium which were essentially the same, thus indicating that, by this time, the airborne radioactive materials had been uniformly mixed in the atmosphere. The concentration was being reduced by half every five days due to deposition and washout. It is interesting to note that no radioactivity from Chernobyl was observed at monitoring stations in the Southern Hemisphere, in Botswana, Australia and New Zealand, thus indicating that the contamination from Chernobyl did not spread across the equator.

Garland and Cambray (in press) assumed that, by 14 May the caesium was uniformly distributed throughout the Northern Hemisphere troposphere and by further assuming that the rate of depletion measured from 14 May also held before that date, they were able to arrive at an estimate of the total release of caesium 137 from the accident of about 1.3 million curies. While they ascribe an uncertainty of a factor of about four to this estimate, it nevertheless agrees remarkably well with other estimates obtained by totally different methods.

The quantity of caesium deposited in the same area during 1963 as a result of nuclear testing has been estimated as four million curies and the Chernobyl caesium fallout is not more than one tenth of the 24 million curies of caesium 137 estimated to have been deposited as a result of all weapons tests. This is of some importance in view of the statement, widely reported in the press, by some "environmentalists" that the emissions from Chernobyl were greater than all of the emissions from weapons testing. That statement is patently untrue.

Chapter Six

The Dawn of the Nuclear Age

In the 1930s there was considerable interest in the popular press about the activities of physicists, who, it was suggested, were on the brink of "splitting the atom." The event, when it occurred, would herald the dawn of a new age in which unlimited supplies of energy would be available at little or no cost. To understand what was going on it is necessary to know something about what atoms are and what they are made of. This is explained in Insert 6. The atoms of different elements are different, but each consists of a positively charged nucleus of protons and neutrons surrounded by a solar system of negatively charged electrons in planetary orbit "shells."

The atom, like the solar system, is mainly empty space. When two atoms approach one another it is the planetary electrons which meet first of all. If an atom with one electron in its outer orbit comes in contact with another atom, of a different element, with one space in its outer orbit, they may combine to make a molecule of a compound. In this way a sodium atom may combine with a chlorine atom to make a molecule of sodium chloride. This in itself is interesting, for sodium is a very reactive metal which has to be kept in oil, as it would otherwise catch fire by reaction with the moisture in the atmosphere, and chlorine is a poisonous gas. However the compound formed is known to us as the innocuous sodium chloride or common salt which has a place in everyone's kitchen. In chemical shorthand this reaction is described by an equation:

$$Na + Cl = NaCl$$

which just means that one atom of sodium reacts with one atom of chlorine to form one molecule of sodium chloride.

Insert 6

What Is the World Made Of?

Everything on earth is made of just two kinds of substance, elements and compounds. An element is a substance which cannot be made into anything simpler by chemical means. Copper, iron, oxygen, sulphur, sodium, hydrogen and sodium are all elements and there are about a hundred of them altogether. When two or more elements combine chemically they form a compound. Sodium chloride (common salt), sulphur dioxide and water (hydrogen oxide) are all compounds.

If you take a piece of an element, say copper, and cut it in two you will have two bits of copper. If you then take one of those bits and cut that in two and keep on doing that you will eventually end up with a piece of copper so small that if you cut it in two again you would no longer have two even smaller bits of copper but something else. At that stage you would have got down to an *atom* of copper. An atom is the smallest part of an element that can lead an independent existence. Two or more atoms joined together form a *molecule*. Note that while you can have a molecule of an element or of a compound, you cannot have an atom of a compound. The atom itself is made up of even smaller building blocks. It is like a miniature solar system with a nucleus (the "sun") carrying a positive electrical charge and surrounded by negatively charged electrons moving in orbits, just like the planets. The nucleus itself is made up of protons (positively charged) and neutrons (uncharged). It is the number of protons in the nucleus (and the equal number of electrons in orbit) that determines the chemical nature of the atom. The hydrogen atom has one proton and one electron, the helium atom two protons and two electrons and so on. The uranium 235 atom has 92 protons and 143 neutrons in its nucleus and 92 planetary electrons.

A more familiar reaction is one that occurs everytime we burn wood or coal or light a barbeque. This is the burning reaction of which the chemical equation is as follows:

$$C + O_2 = CO_2$$

which says that one atom of carbon reacts with two atoms (or one molecule) of oxygen to form one molecule of carbon dioxide. This equation is not quite complete for none of us burns wood in order to get carbon dioxide, we do it for the heat given off in the reaction. The complete equation reads:

$$C + O_2 = CO_2 + 93{,}000 \text{ calories per mole.}$$

A mole is the molecular weight of carbon in grams, or 12 grams. A calorie is the quantity of heat required to raise the temperature of one gram of water by one degree Celsius, so that 12 grams (or rather less than half an ounce) of carbon will provide enough heat to raise the temperature of 93 litres of water by 1 degree Celsius or one litre of water from 7 degrees Celsius to boiling. It is on this reaction that all of our consumption of fossil fuel ultimately depends. The carbon dioxide is of no interest to us and escapes to the atmosphere where, as will be seen, it is storing up problems for the future. It is of some interest to calculate how much energy is liberated in the basic burning reaction for every atom of carbon involved. There are six hundred thousand billion billion atoms (6×10^{23}) in twelve grams of carbon and when we divide that into 93,000 calories we get an impossibly small number of calories. Physicists respond to that simply by inventing a new unit of energy, the electron volt, which is the quantity of energy acquired by an electron when it is accelerated through a potential difference of one volt. That is a very small quantity of energy and we can calculate that each atom of carbon involved in the burning reaction liberates 4 electron volts of energy. This is fairly typical of the quantities of energy liberated per atom in chemical reactions.

Nuclear Reactions

The excitement generated by the possibility of splitting the atom was due to the fact that there seemed to be the chance of getting very much more energy from a reaction involving the nucleus of the atom than from one involving (as chemical reactions do) the planetary electrons. For example in 1932 J. D. Cockcroft and E. T. S. Walton carried out an experiment at the Cavendish Laboratory in Cambridge in which they accelerated protons through a potential difference of half a million volts and "fired" them at a target of lithium atoms. The reaction they observed is described by the following equation:

$$_{3}Li^{7} + {}_{1}H^{1} \longrightarrow {}_{2}He^{4} + {}_{2}He^{4}$$

This merely says that lithium atoms (containing 3 protons and 4 neutrons in their nucleus to give a total mass of 7) when bombarded by protons "split" to give two helium atoms (each with 2 protons and 2 neutrons). Everything about this equation adds up; they started off with four protons and four neutrons on the left-hand side and ended up with four protons and four neutrons on the right-hand side. The only remarkable thing is that in this "splitting of the atom" a comparatively large quantity of energy (compared at any rate to that released in a chemcial reaction), 17.2 million electron volts, is released. Where does this energy come from? A careful study of the masses of the reacting particles and the products of the reaction showed that a slight amount of mass had apparently disappeared. The amount of energy liberated coincided exactly with that which would have been forecast using Einstein's equation $E = mc^2$ for the equivalence of mass and energy, (E is the energy equivalent of a mass m and c is the velocity of light). Since this experiment in 1932 (for which Cockcroft and Walton were awarded the Nobel Prize in Physics) many nuclear reactions have been studied in detail and, in every case, the quantity of energy released agrees, within the limits of the experimental errors, with the change of mass accompanying the process — a convincing confirmation of the validity of the Einstein equation.

During the ensuing few years many other nuclear reactions were performed and they were years of great excitement for the physicists who were uncovering the secrets of the atom at an unparalleled rate. The physicists however did not share the optimism of the popular press as regards the wholesale generation of cheap energy. As Lord Rutherford pointed out, while the energy released in a single nuclear reaction was large, only a very small proportion of the protons used, for example, in the proton-lithium reaction, found their target — about one in a billion — so that overall the entire experiment consumed large quantities of energy rather than generating them. He said that the chances of causing a reaction with a single proton were about the same as that of a blindfolded man hitting a single mosquito with a rifle shot in the Royal Albert Hall and concluded that "the outlook for gaining useful energy by artificial processes of transformation does not look promising." Most physicists were of the same opinion, but they were shortly to change their minds.

The Neutron

In 1932 Chadwick had discovered the existence of the neutron, a particle of similar mass to the proton, but with no electric charge. This raised the possibility of a whole new range of nuclear reactions, for the neutron, being uncharged, could approach the nuclei of atoms without being repelled by the positive charge on the nucleus. It was not long before physicists everywhere were bombarding every atom imaginable with neutrons. The results were mostly predictable. Consider the simplest case in which atoms of hydrogen are bombarded by neutrons; some of the neutrons hit hydrogen nuclei and are absorbed by them (Figure 12). This doubles the mass of the nucleus, but does not alter the chemical nature of the atom since this is determined by the number of electrons in orbit which is the same as the number of protons in the nucleus. Thus we still have an atom of hydrogen, but it is a different sort of hydrogen atom since its mass has been doubled. Thus a new *isotope* of hydrogen, heavy hydrogen or deuterium has been created.

We can do this again by hitting a deuterium nucleus with another neutron. We end up with a hydrogen atom three times as massive as the original, its nucleus containing one proton (as do all hydrogen atoms) and two neutrons, which is yet another isotope of hydrogen — tritium. It turns out that the nuclear configuration of two neutrons and a proton is rather unstable and so the tritium atoms are continually trying to achieve a more stable structure. In other words they are radioactive and eventually decay by emitting an electron from the nucleus. But there aren't any electrons in the nucleus! Where does the electron come from?

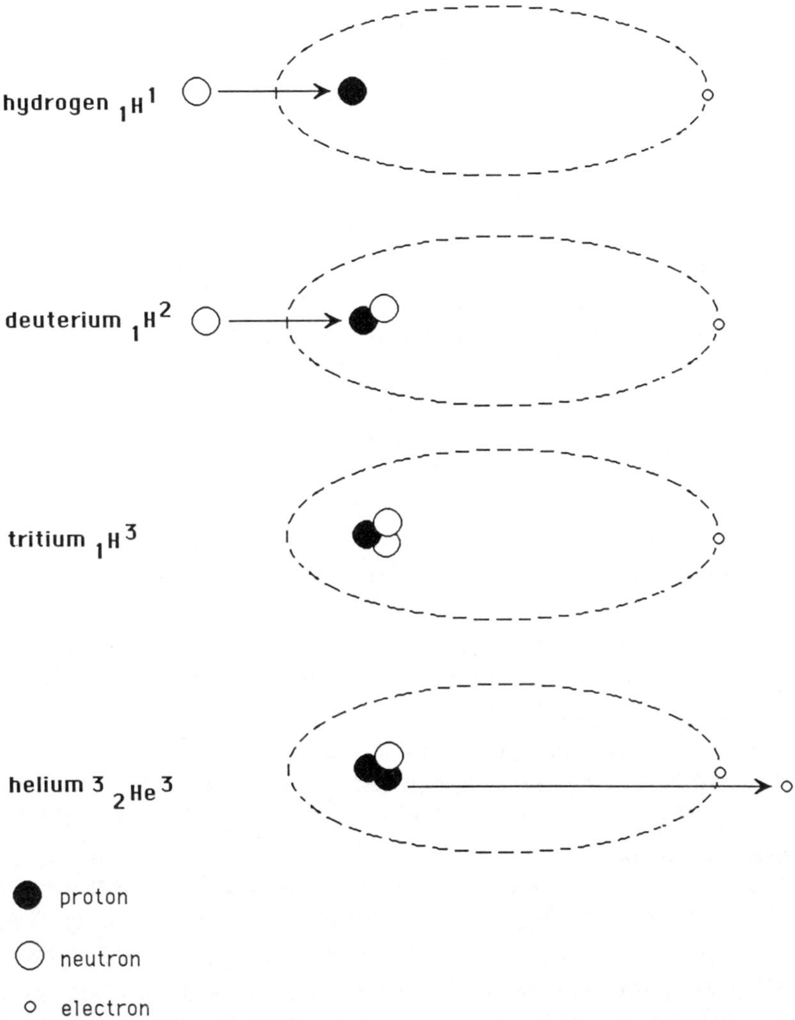

Figure 12 Transmutation of Hydrogen into Deuterium and Tritium by Neutron Bombardment and Decay of Tritium to Helium 3

Figure 12 gives the answer; in effect one of the neutrons in the nucleus changes into a proton and an electron, the electron is emitted and the proton remains. But now we have done something quite fundamental, we have changed the chemical nature of the hydrogen atom. It no longer has one proton in its nucleus, it has two and is therefore no longer a hydrogen atom, but a helium atom. Thus an atom of one element has been changed into an atom of another. Nuclei which have absorbed an additional neutron and been made unstable can "decay" to a more stable form in other ways too. They can emit gamma radiation (rather like X-rays, but more penetrating) or alpha particles (the nuclei of the helium atom consisting of two protons and two neutrons).

In all of the cases in which an atom of one element was changed into something else by neutron bombardment the final result was either another isotope of the same element (if gamma radiation was emitted), an isotope of the element with atomic number one more than the parent as in the tritium case (if an electron was emitted) or an isotope of an element with atomic number two less than the parent (if an alpha particle was emitted). The properties of these resulting atoms could thus be predicted and recognised from the known properties of other elements. Imagine the surprise therefore when uranium was bombarded with neutrons and, instead of one atom with properties of one of uranium's near neighbours, two atoms appeared whose properties were similar to elements with about half the atomic weight of uranium.

It took several years for the true explanation to appear. On 16 January 1939 Lise Meitner and Otto Frisch wrote, in a letter to Nature, "It seems possible that the uranium nucleus . . . may, after neutron capture, divide itself into two nuclei of roughly equal size." This process was called fission. It was noteworthy for two reasons, first of all, the amount of energy liberated in fission was approximately 200 million electron volts (Mev) per atom of uranium, an amount very much greater than observed in any other nuclear reaction. More important however was the fact that each atom of uranium which fissioned appeared to release two or three neutrons in addition to producing the energy and the fission products. If things were suitably arranged it seemed that it might be possible for these additional neutrons to cause more fissions in other uranium atoms and the possibility of a self-sustaining chain reaction arose.

The Fission Process

In nature uranium occurs as a mixture of the two isotopes uranium 235 ($_{92}U^{235}$) and uranium 238 ($_{92}U^{238}$). About 0.7 per cent of natural uranium consists of uranium 235. It turns out that fission can be caused in uranium by both "slow" (or thermal) neutrons, of energy around 0.03 electron volts, or by "fast" neutrons of energy 1 million electron volts. Uranium 238 is fissionable only by fast neutrons but both types will cause fisson in uranium 235. In addition, uranium 238 has the property of absorbing neutrons (without causing fission) if they have energies of about 5 electron volts. This absorption turns out to be quite important, for any neutrons of this energy which happen to meet a uranium 238 atom will be absorbed and thus prevented from causing further fissions. Because of this absorption it is not possible to sustain a chain reaction in pure uranium 238.

Let's suppose we have a lump of natural uranium and a stray neutron arrives from somewhere and hits the nucleus of an atom of uranium 235 causing fission. As a result of this say three more neutrons are emitted. One may escape out of the uranium and be lost, another may be slowed down to an energy of 5 electron volts and hit a uranium 238 nucleus and be absorbed, the third one may cause another fission in a uranium 235 atom (Figure 13). In order to achieve a self-sustaining chain reaction we have somehow to juggle these numbers so that the number of neutrons causing further fissions is one, or greater. If we can do this the chain reaction will spread with astonishing rapidity.

If, for example, the number of neutrons produced in each fission which cause other fissions is 2, then, starting with a single neutron causing a fission, successive generations would cause 2, 4, 8, 16, 32, 64, 128 and so on fissions. In eighty generations there would be more than 10^{24} fissions occurring — sufficient to cause the fission of every uranium nucleus in 240 grams of

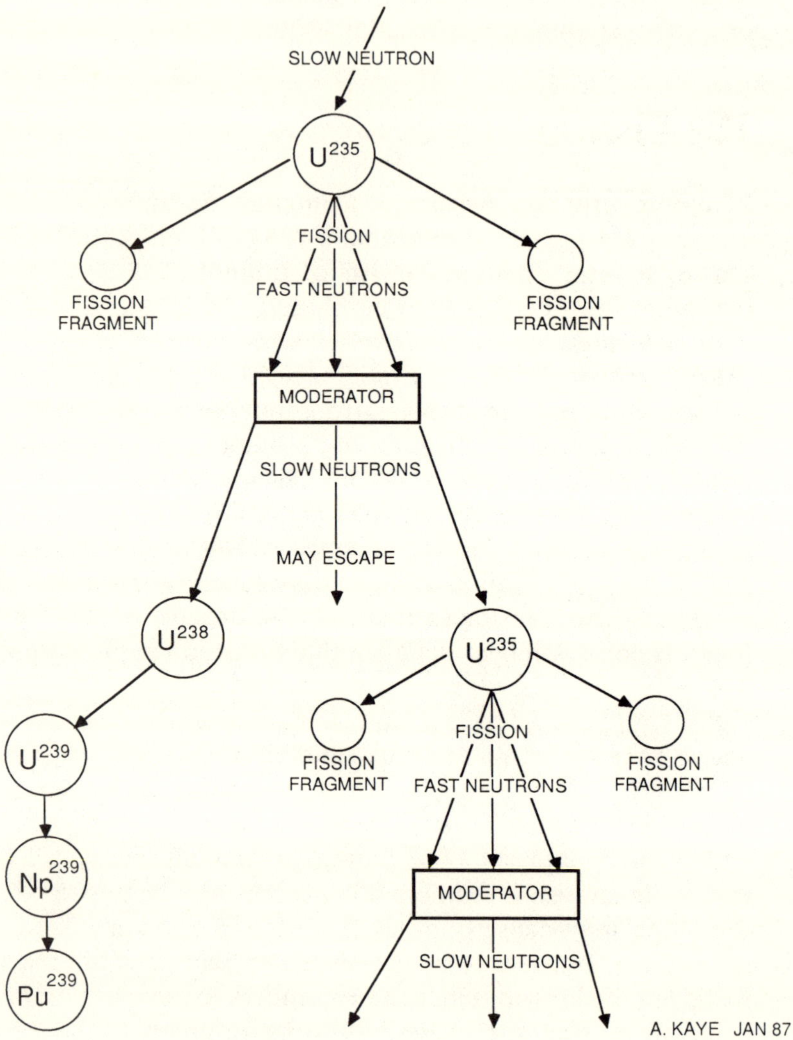

Figure 13 The Fission Process

uranium. The total energy released would be more than five million kilowatt hours and this would be released in less than a millionth of a second. The result would be an enormous explosion. It was this possibility which spurred research both among the Allies and Axis powers during World War II and culminated in the atomic bomb explosions at Hiroshima and Nagasaki in 1945.

The possibility of causing a chain reaction therefore depends on several neutrons being emitted as a result of each fission. We must ensure however that not too many of these neutrons escape from the uranium itself without causing other fissions. If we start off with a small piece of uranium it is possible that many of the neutrons produced in fission will get out of the uranium without encountering another uranium nucleus. However, as we increase the size of the uranium, this possibility gets less and less and eventually we arrive at a size where further fissions will occur spontaneously — we have reached a "critical mass" of uranium, so called not because it is necessarily dangerous, but because it is the minimum size which will sustain a chain reaction.

There is also the possibility that some of the neutrons will have just the right energy to be absorbed by uranium 238 nuclei (and remember that there are 139 times as many of those in natural uranium as there are uranium 235 nuclei). Such a neutron will be "lost" as far as producing further fissions is concerned, but starts off a series of reactions which is also quite important as far as nuclear energy is concerned. First of all an atom of uranium 239 is formed:

$$_{92}U^{238} + n \text{ ———> } _{92}U^{239}$$

Uranium 239 is radioactive and decays, with a half life of 23 minutes, to form neptunium 239 emitting an electron:

$$_{92}U^{239} \text{ ———> } _{93}Np^{239} + \text{electron}$$

Neptunium 239 is also unstable and decays, again by beta decay, with a half life of 2.3 days to form plutonium 239:

$$_{93}Np^{239} \longrightarrow {}_{94}Pu^{239} + \text{electron}$$

Plutonium has no stable isotopes and the half life of plutonium 239 is 24,000 years. This is so short, in relation to the age of the earth, that any plutonium which was present when the earth was formed has long since decayed away. Although trace quantities of plutonium are continually formed by cosmic ray bombardment, it is mostly a man-made element and its importance in nuclear energy is that it also is fissionable by fast neutrons.

The Atomic Bomb

To make an atomic bomb it was necessary either to have nearly pure uranium 235 or plutonium 239. In the event three bombs were made in the first half of 1945, two from plutonium and one from uranium 235. One of the plutonium bombs was used in the test at Alamagordo, New Mexico on 16 July 1945, the uranium bomb was dropped on Nagasaki on 6 August and the second plutonium bomb on Hiroshima on 9 August. The uranium 235 was separated from natural uranium by gaseous diffusion. First of all natural uranium was chemically combined with fluorine to form the gaseous compound uranium hexafluoride. This gas was then made to diffuse through a series of membranes penetrated by very fine holes. The rate at which a gas diffuses depends on its molecular weight and there is a slight difference between the molecular weight of a uranium hexafluoride molecule containing an atom of uranium 235 and one containing an atom of uranium 238. The former will diffuse slightly faster, so that the gas penetrating the membrane will be slightly enriched in uranium 235. The enrichment factor is only 1.0014 and this means that it takes about 4000 diffusion stages to raise the uranium 235 content of natural uranium from 0.7 per cent to 99 per cent. The diffusion plant built at Oak Ridge, Tennessee was enormous.

The obtaining of appreciable quantities of plutonium 239 was quite a different problem. The only way of doing so was to maintain a chain reaction in natural uranium. It was necessary to use natural uranium because the uranium 235 isotope was essential to maintain the chain reaction and the uranium 238 was needed to absorb some of the neutrons emitted in fission so that plutonium could be formed and eventually separated chemically from the uranium.

The chain reaction had to be controllable, the fast neutrons produced in fission had to be slowed down so that some of them (when their energy was reduced to 5 electron volts), could be absorbed by the uranium 238 and the remainder slowed down still further, to 0.03 electron volts, to cause further fissions in uranium 235 and maintain the chain reaction. The neutrons are slowed down by causing them to collide with other atoms, but it is necessary to choose these "moderating" atoms rather carefully so that they slow down the neutrons but do not absorb them. It was discovered that the best substances to do this were hydrogen, deuterium, helium, beryllium, carbon and oxygen. A substance like water, made up of hydrogen and oxygen would seem to be a cheap and convenient moderator, but in fact hydrogen absorbs too many neutrons to be effective as a moderator unless the uranium is artificially enriched in uranium 235. Heavy water or deuterium oxide and carbon both make good moderators and, carbon in the form of graphite was used in the first "reactors," mainly because of its cheapness and ready availability.

The First Nuclear Reactor

The first nuclear reactor was assembled in a squash court under the West Stands of Stagg Field at the University of Chicago in December 1942. The plan was to build a cubic lattice of lumps of uranium and uranium oxide within a sphere of graphite. The graphite was cut in bricks and built up in layers, alternate layers containing lumps of uranium or uranium oxide. Since the structure was made by piling one layer on top of another, it was called a pile. While it was being built cadmium strips were inserted as a safety measure. Cadmium is a very effective neutron absorber and it was thought that without them, as the pile approached its critical size a neutron from cosmic rays or a spontaneous fission might accidentally set off the chain reaction.

As it turned out, this was a wise precaution since the critical state was reached much sooner than was anticipated. As the cadmium strips were cautiously withdrawn the neutron intensity within the pile was measured. On the afternoon of 2 December 1942 a sharp increase in the neutron intensity within the pile showed that a self sustaining chain reaction had been achieved. The pile was first operated at a power of 0.5 watts, but on 12 December 1942 it was raised to 200 watts. This was certainly a small beginning and a long way from the 1000 million watts (electrical) achieved by the Chernobyl reactor. The Chicago reactor was, of course, a very rudimentary reactor indeed and its only purpose was to determine whether the chain reaction fission process was feasible. In this it was brilliantly successful and it was immediately followed by the construction of large reactors at Hanford (Washington) for the full-scale production of plutonium.

A great deal of controversy and argument (not least among scientists themselves) has followed the construction and use of the atomic bombs. J. R. Oppenheimer, who was in charge of the design and assembly of the actual bombs at Los Alamos remarked "In some sort of crude sense which no vulgarity, no humor, no over statement can quite extinguish, the physicists have known sin, and this is a knowledge they cannot lose." This implied that there was some sin involved in developing the bomb and, especially, in using it.

It is easy to be wise after the event, but a dispassionate observer might find it difficult to perceive that the Allied powers had any choice. Scientists on the Axis side were just as aware of the potential of nuclear fission and intelligence sources showed that the production of the atomic bomb was being actively pursued in Germany. It has been argued that Japan was on the brink of collapse at the beginning of August 1945 but one physicist (the author), at the time in the Army in the Far East, saw precious little evidence of that. The casualties suffered by the American forces in their inch-by-inch re-conquest of the

Pacific Islands left no hope for optimism that the final conquest of Japan would be easy.

The atomic bomb was undoubtedly a terrible weapon. One should remember the context in which the decision to use it was taken. Those were the days of the gas chambers and the Burma-Siam Railway. That decision may have saved many thousands of lives and possibly prevented a third world war. The current cold war truce, uneasy though it may be, seems preferable to what might have ensued without nuclear weapons, which are, after all, the only weapons in the history of mankind which have been used on only two occasions, and by only one nation.

Peaceful Uses of Nuclear Energy

During the development of atomic weapons the whole project was shrouded in secrecy and this attitude spilled over into peacetime. In fact there had been leakages of information to Soviet agents during World War II and when President Truman broke the news to Mr. Stalin at the Potsdam Conference on 24 July 1945 "that we had a new weapon of unusual destructive force," Stalin's reply was unexpectedly brief. He was glad to hear it and hoped we would make "good use of it against the Japanese." Truman was puzzled by this cool reception from Stalin, but by that time Klaus Fuchs and other agents had provided the Russians with a very detailed picture of the weapon that had recently been tested.

The Soviets had already been working on a fission weapon of their own for almost two years, and carried out their first atmospheric test of a fission weapon in September 1949. In the meantime the race was on in Britain and France to develop their own nuclear weapons. Britain carried out tests in Australia at the Montebello Islands in 1952 and at Maralinga in 1954. Chickens have a way of coming home to roost and in 1984 the Australian government set up a Royal Commission to investigate the less-than-perfect conduct of the British tests as far as safety precautions and health physics were concerned. However secret the major powers endeavoured to keep their weapons research and development there was a new atmosphere which was advocating the development of nuclear energy for peaceful purposes.

It was in the early 1950s that the serious development of power-producing reactors began. It was not unreasonable that they should be based on the experience that had been built up in the military programmes. All of the reactors used thermal neutrons for the chain reaction and this meant that the fuel was uranium 235 whether contained in natural or enriched uranium. The purpose of the reactors was to generate electricity and it is important to remember that the only difference between a nuclear generating station and a fossil fuel station is the means employed to heat the water and make steam (Figure 14). In the coal- or oil-fired stations the fuel is burned to heat water and raise steam. The steam is then passed to a steam turbine which in turn rotates the alternator which generates the electricity. In the nuclear station the energy from the fission of uranium is used to heat water and raise steam and, from then on, the two types of station are identical.

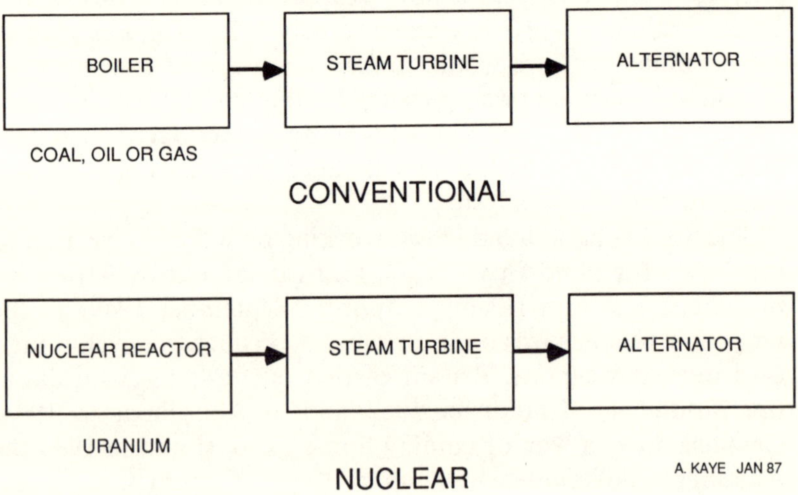

Figure 14 Conventional and Nuclear Electricity Generating Stations

It was fairly obvious in the early fifties that the first power-generating reactors would have to use thermal neutrons, since the technology of "fast" reactors had not been studied at all at that time. There were then two major decisions to be made, what moderator to use and what means of cooling to employ. The early reactors had been (with one exception, a graphite-moderated, water-cooled reactor at Hanford, Washington) graphite-moderated and air-cooled. However these reactors were all designed primarily for the production of plutonium for use in weapons and thermal efficiency was not a major requirement.

Even conventional electrical generating stations are rather poor converters of heat into electricity at the best of times. While the efficiency of the boiler can be greater than 80 per cent and the alternator more than 99 per cent, steam turbines have, for thermodynamic reasons, a maximum theoretical efficiency (depending on the temperature at which steam is discharged from the turbine) of about 50 percent. The combination of these factors brings the overall efficiency of even the best conventional stations down to about 40 per cent. The efficiency can be raised only by increasing the temperature of the steam entering the turbine and there are engineering limitations on this.

In the nuclear reactor safety and neutron economy considerations place even more stringent constraints on the materials which can be used and these, in turn, limit the temperature at which steam can be produced even more than in conventional stations. Since there was no practical experience of using reactors as a prime heat source for generating electricity the choice had to be made, on rather inadequate data, between the various systems. As has been seen, the design of a reactor is conceptually simple. It is necessary to intersperse fuel with moderator together with some sort of neutron-absorbing control rods, which can be inserted or withdrawn as desired. When a critical assembly is reached the control rods are withdrawn gradually and the power level of the reactor rises to the desired level when the control rod withdrawal is stopped.

There are, however, some complications. In a reactor operating at a power of 1000 megawatts (heat) 31 billion billion (3.1×10^{19}) atoms of uranium 235 are undergoing fission every second. Each of these fissions results in the formation of two fission-product atoms. The range of the fission products that have been detected experimentally is from mass number 72, an isotope of zinc (atomic number 30) to mass number 158, believed to be an isotope of europium (atomic number 63). Most of the 34 different elements in this range are produced directly in fission, and about 97 per cent of the uranium 235 atoms undergoing fission yield products which fall into two groups a "light" group with mass numbers from 85 to 104 and a "heavy" group with mass numbers from 130 to 149.

The most probable type of fission, representing about 6.5 per cent of the total gives products with mass numbers of 95 and 139. Most, if not all, of these fission products are radioactive and decay with the emission of electrons and gamma rays, with half lives ranging from the very short (a few thousandths of a second) to the very long (around a million years). The fission products are formed within the reactor fuel and, if they were not prevented from doing so, some of them might diffuse out of the fuel and contaminate the reactor. To prevent this happening the fuel elements have to be hermetically sealed in cans. The can material chosen must maintain its integrity indefinitely at the working temperature of the reactor (and indeed, at temperatures much above normal operating temperature, to take care of possible fault conditions), and also, it must not be a significant absorber of neutrons. This imposes stringent constraints on the can material and the choice of can materials for a reactor fuelled by natural uranium is really limited to certain aluminium, magnesium and zirconium alloys. Stainless steel, otherwise a practically ideal can material is such an effective neutron absorber that reactor fuel has to be considerably enriched in uranium 235.

Countries wishing to enter the business of nuclear generation of electricity had therefore to make the basic decision of what moderator to use, what coolant to use and what can material to use for the fuel elements. Some hard choices had to be made, for it was impracticable, on the grounds of cost and resources, to back all the horses in the race. In the event different countries made entirely different choices as is shown in Table 6. Each of these programmes has been successful to some degree and the different types of reactors resulting from them and more especially their advantages and disadvantages are discussed in Chapter 7.

Table 6

Types of Reactors in Civil Power Programmes

Britain	Initially "Magnox"-type reactors, graphite-moderated, carbon-dioxide-cooled, natural uranium metal fuel canned in magnesium alloy. Advanced gas-cooled Reactors, (AGR) graphite-moderated carbon-dioxide-cooled, enriched uranium oxide fuel canned in stainless steel.
Canada	"Candu" reactors, heavy-water-moderated, heavy-water-cooled, natural uranium oxide fuel canned in zirconium alloy.
France	Initially gas-cooled reactors, similar to the British, now switched to light-water reactors and fast reactors.
Germany	Light-water reactors.
Italy	Initially bought a Magnox reactor, now light water.
Japan	Initially bought a Magnox reactor, now light water.
US	Pressurised and boiling-water reactors, enriched oxide fuel canned in Zircaloy*, light-water-moderated, light-water-cooled.

*a zirconium alloy

Nuclear Fuel Cycles

In a coal-fired generating station the coal is burned, more coal is added and the ash is removed. Things are more complex in a nuclear plant for, as soon as the first fission takes place changes begin to occur in the fuel which eventually reduce the neutron economy so that the fuel has to be changed long before all of the uranium 235 has undergone fission. Every atom of uranium that undergoes fission produces two atoms of fission products. Some of these absorb neutrons which could otherwise be used for more fissions. Some of the neutrons are themselves absorbed by uranium 238 atoms (of which there are 139 times as many of these as there are of uranium 235). As shown in Figure 13 these uranium 238 atoms will eventually end up as atoms of plutonium 239 and these will also absorb and (when some of them fission) also produce some neutrons.

At some stage therefore the fuel will have to be changed or it will no longer be able to maintain a chain reaction. It is essential, however, to change it long before that and it is the usual practice to change the fuel more or less continuously so that the reactor can be run indefinitely at full power. This can be accomplished either by shutting the reactor down periodically and changing some fraction (perhaps a third or a quarter) of the fuel at each shutdown or by designing the reactor so that the fuel can be changed while the reactor is running at full power.

The latter method is obviously preferable in that it reduces the "down time" of the reactor (for shutting the reactor down for a fuel change and starting it up afterwards is a time-consuming operation), but it does complicate the design of the reactor, for arrangements have to be made to withdraw highly radioactive fuel elements into some sort of mobile shielded container while the reactor is running. This is complicated by the fact that in many power reactors the coolant is under considerable pressure and a successful fuel-charging machine is a very sophisticated piece of engineering. When the fuel is removed from the reactor it is intensely radioactive. Many of the radioactive fission products have very short half lives and so the initial radioactive decay is quite rapid. Eventually only the longer-lived fission products remain and the rate of decay slows down. Because of the heat generated by the decay of fisson products in the fuel removed from the reactor, cooling is essential. Even the fuel-recharging machines have to have provision made for cooling the fuel elements removed from the reactor for the short time before the elements are transferred to storage.

The word "spent" is a misnomer for the fuel elements removed from the reactor, for only a proportion of the uranium 235 contained in the fuel rod has undergone fission. The higher the proportion of the uranium 235 which is used up in the reactor, the better the economics of the fuel cycle. In addition the fuel element will contain plutonium 239 formed by the absorption of neutrons by the uranium 238 which constitutes 99.3 per cent of the natural uranium.

It is possible to retrieve the unspent uranium from the fuel element together with the plutonium by carrying out a chemical separation on the used fuel element. Because they are highly radioactive the separation has to be done by remote control in an automated and heavily shielded chemical plant. Such facilities exist in a number of countries, for example, Britain, France, Russia and the United States. These chemical separation plants were originally set up to process used reactor fuel so that the plutonium could be extracted for use in nuclear weapons. It has not in the past, been worthwhile for countries (such as Japan) operating only a few such reactors to build chemical processing factories so they enter into contracts with the others for the reprocessing of their fuel. Canada, which has a considerable nuclear programme makes no attempt to process spent fuel elements and stores them under water. In a way this is just like having money in the bank for, if the need should ever arise (through scarcity of natural uranium, or the emergence of a Canadian fast reactor programme) to reprocess that material, it will always be there. The chemical separation of used nuclear fuel inevitably results in the accumulation of quantities of highly radioactive waste. It is necessary to provide safe and reliable storage for these for long periods of time, and it is this aspect of nuclear energy which has probably caused most disquiet among concerned members of environmental action groups and, indeed, the general public at least up to the Chernobyl accident. It is also the main area of nuclear generation of electricity in which the problems have not as yet been solved to the satisfaction of even its most enthusiastic protagonists. The methods employed and proposed for nuclear waste storage will be discussed in Chapter Ten.

Chapter Seven

Reactors for Energy Generation

In the early 1950s the "nuclear powers" could be counted on the fingers of one hand — the USA, Russia, Britain, France and Canada. The way ahead was far from clear. Of the five only Canada had a purely civil interest in nuclear energy, the remainder wanted to manufacture plutonium for weapons as well as to generate electricity. This limits the options to some degree for reactors can be designed primarily for the production of plutonium, or for the production of energy or for a rather unsatisfactory compromise. We shall limit the discussion here to civil reactors.

All nuclear reactors consist of a core in which uranium 235 undergoes fission and a moderator in which the fast neutrons emitted in fission are slowed down so that they may cause other fissions. The heat generated in the core is carried away by a coolant which transfers this heat (in a heat exchanger) to water in a secondary circuit. This water boils and the energy in the resulting steam is changed to rotational energy in a turbine. This is illustrated in Figure 15 which shows the features common to all reactors. The various types differ in the layout of these features, the canning material, the type of moderator, the coolant, the pressure under which the coolant operates and whether the reactor uses natural or enriched uranium as fuel. There are thus many different possible types of reactor, but only a few of them have come into common use. These will now be described.

Light-Water Reactors

The USA chose light-water-moderated and cooled reactors

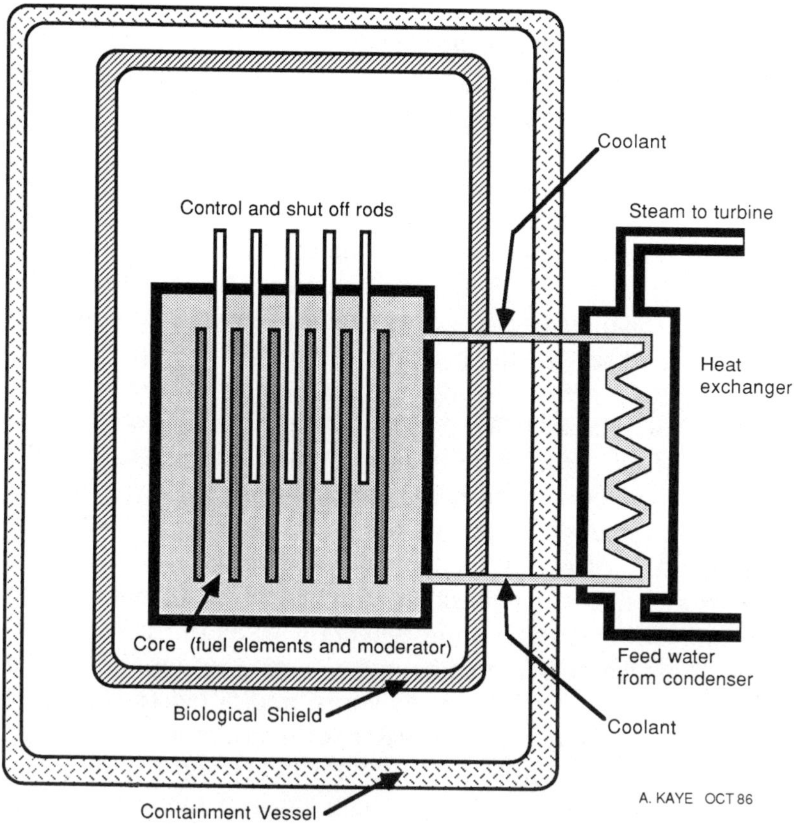

Figure 15 Features Common to All Reactors

which required the use of uranium considerably enriched in uranium 235. This was not a great problem for there was already a large fuel-enrichment capacity built for weapons production. The disadvantage of the system is that it makes the fuel cycle more expensive; the advantages are that it makes for a compact core which may lead to easier export sales and, once the sales have been made, since most countries buying reactors do not have fuel-enrichment capacity, to the possibility of future business in the reprocessing and enrichment of fuel. The compact core also implies a high power density, i.e., a lot of energy is generated in a small space and this may lead to control problems. The first pressurised water reactor was designed as a power unit for the submarine *USS Nautilus* which was launched in 1954 and in 1957 a land-based version was installed at Shippingport, Pennsylvania and became the first nuclear power station in the USA.

The fuel of the PWR is uranium dioxide, UO_2, in cylindrical pellets about a centimetre in diameter contained in a zirconium alloy ("zircaloy") tube nearly 3 metres long. The reactor is both moderated and cooled by ordinary "light" water and because of the high absorption of neutrons by water the fuel has to be enriched so that the uranium 235 content is 3 per cent instead of the 0.7 per cent of natural uranium. The reactor core is contained in a thick-walled pressure vessel of welded steel whose lid is held down by a circle of heavy bolts. The pressure of the water in the reactor is about 150 atmospheres and it acts both as moderator and coolant. At the top of the core the hot water, at a temperature of about 320°C is piped through the side of the pressure vessel to a heat exchanger in which it is forced through thousands of small pipes immersed in the water of the secondary circuit. This water is heated by contact and, because it is under much lower pressure it turns to steam and is fed to the steam turbine. The primary coolant water is then pumped back to the reactor to be heated again. A second primary coolant loop contains a "pressure equaliser" in which some of the water will be allowed to evaporate or condense in order to maintain the coolant at the correct pressure as the reactor power level is changed. The control and shut-off rods are generally suspended inside the pressure vessel, above the reactor core and they are controlled through the lid of the pressure vessel.

The reactor is refuelled only when it is shut down. It is first allowed to cool down and then a large tank on top of the core is filled with water to provide shielding and cooling water to remove the fission-product heat. The lid of the pressure vessel is then removed and the fuel replaced. Because the process takes so long a fairly large fraction (about one third) of the fuel is replaced at each shut-down and the reactor is generally shut down for refuelling once per year. A typical PWR contains about 100 tonnes of 3 per cent enriched UO_2 in zirconium alloy cans in a core about three metres in diameter and three metres high. Its pressure vessel is made of welded steel 20 centimetres thick and the whole core and primary coolant circuit is contained in a biological shield of concrete about two metres thick. The reactor is also provided with an emergency cooling system which is designed to take over and provide adequate cooling should any malfunction occur in the cooling system. Although the power density is high (about 100 kilowatts per litre volume) the need for keeping the coolant pressure down to 150 atmospheres means that the maximum coolant temperature is only about 320°C. The overall efficiency attained is thus not very high and the thermal efficiency (the ratio of electrical energy produced to total heat produced) is only about 32 per cent.

In the boiling water reactor (BWR) the water is allowed to boil within in the core and the resulting steam is piped directly to the turbine. There is thus no secondary coolant circuit as with the PWR. Provision has to be made for dealing with the steam should any malfunction render the turbine unable to accept it. The reactor is therefore enclosed inside a secondary container, a reinforced concrete vessel, also the biological shield, from which steam will be channelled into a pool of water under the reactor. The BWR is also refuelled when shut down. There are more light water reactors operating throughout the world than any other type.

The Russian version of the Pressurised Water Reactor is the VVER which, it is thought will be the main reactor type for the future expansion of Soviet nuclear power. The first stage of its development was a 220Mw(e) station built at Novovoronezh in 1964. The design was then standardised at 440Mw(e) and exported to Finland (Loviisa 1 and 2) and other countries. This is the only type of Soviet reactor which has been exported. The next stage in the development was to increase output the 1000Mw(e) and the VVER-1000 units were the first in the USSR to have a form of containment. The reactor and its primary circuit are enclosed in a reinforced concrete containment vessel with a diameter of 45 metres. There are now 41 of these VVER-1000 units in operation, under construction or planned on 13 sites in the USSR. Of these five are in operation and the remainder are coming on stream at the rate of one reactor about every three months. With multiple units being constructed on one site a building production line has been adopted which enables a unit to be completed in 5 years. About 20 per cent of Russia's fossil fuel consumption is used for the production of hot water for district home heating. The VVER-1000 stations are used for both electricity and district heating and dual purpose units are under construction at Kharhov, Minsk and Odessa. Reactors for district heating only have also been developed from the VVER design and are being constructed at Voronezh and Gorki, the latter station being only two kilometres from the city boundary.

Graphite-moderated Water-cooled Reactors

The first graphite-moderated water cooled reactors were built at Hanford, Washington for the military production of plutonium and the first two began operating in 1944 and the third in 1945. There is one 860Mw graphite-moderated water-cooled reactor still operating at Hanford, the N reactor which was started up in 1963. It is one of the main reactors for American plutonium production and is now in the 23rd year of what was supposed to be a 20-year lifetime. A safety review was performed on this reactor following the Chernobyl accident and, following that, the Department of Energy, which owns and operates the reactor, announced on 12 December 1986 plans for a six-month $50 million upgrading programme. The Hanford N reactor does not have a containment building but has, instead, a pressurised ventilation and filtration system designed to handle the "maximum credible accident."

The Soviet version of the graphite-moderated, water-cooled reactor is the RBMK which uses low-enriched uranium. It is a direct cycle reactor; water is boiled directly within the core of the reactor and led off to drive the turbogenerator. The reactor does not have a single large pressure vessel to contain both fuel and coolant and the fuel elements are contained instead in more than 1600 separate pressure tubes. Water is allowed to boil in the pressure tubes and produces steam at 284°C and a pressure of 70 atmospheres. The steam contains many water droplets which have to be removed before entering the turbine so the steam is collected from the pressure tubes into four huge drums where it is dried. Each of the drums contains more than 400 welds which have to be pressure tight. The graphite temperature under normal conditions in the RBMK reactors is about 700°C which is above its ignition temperature in air. It is therefore blanketed in an inert mixture of helium and nitrogen.

There has been some criticism in the USSR about the standard of construction of the Soviet nuclear reactors and the implication has been that in the haste of mounting a very large programme very quickly shortcuts have been taken and substandard construction tolerated. In the 27 March 1986 edition of *Ukraine Literature* Ms. Lyubov Kovalevska, who is apparently a senior manager at the Chernobyl Plant wrote that poor workmanship and concrete together with thefts and bureaucratic incompetence "were creating a time bomb and these failures would be repaid over the decades to come." While Ms. Kovalevska's criticisms were directed to the construction of the fifth reactor at the site she pointed out that the same mistakes made in the construction of the first reactor had been carried on to subsequent models.

It had previously been reported that the two Ignalia (Lithuania) RBMK reactors suffered from many of the same material and construction problems reported by Ms. Kovalevska for Chernobyl.

Graphite-moderated Gas-cooled Reactors

The British nuclear power programme started off with gas-cooled, graphite-moderated reactors and has, by and large, stuck with them longer than any other country. There have been moves recently to change to PWRs and a marathon public enquiry is underway (1986) to decide whether a PWR should be sited at Sizewell in Suffolk.

The Magnox Reactors

The original reactors are known as "Magnox" reactors, after

the magnesium alloy in which the fuel is canned. The fuel consists of natural uranium rods and the coolant is carbon dioxide at a pressure of 19 atmospheres. Like all graphite-moderated reactors the core is large, a cylinder nearly 14 metres in diameter and 7.5 metres high, containing more than 300 tonnes of uranium and generating about 250 Mw(e). The whole reactor is encased in a pressure vessel of welded steel 10 centimetres thick, and this is enclosed within a concrete biological shield about two metres thick. Four large ducts carry the hot carbon dioxide out of the reactor building to four heat exchangers, one on each corner of the building. There the CO_2 heats water in thousands of tubes and the resulting steam is transferred to the turbine.

The Magnox reactors are refuelled "on load" by moving the charge/discharge machine over an access port on the reactor cap, clamping it down and pressurising it. The shielding plug is removed and a "fuel removing tool," rather like a very long-nosed pair of pliers, pushed down into the core of the reactor where it grabs and lifts a fuel element into the discharge machine which is fitted with a cooling circuit. The "charge" part of the machine is then moved over the access port and inserts a new fuel element in the reactor. The shielding plug is then reinserted, the machine unclamped and the highly radioactive used fuel element transported to the "cooling pond" where it is stored under five metres of water for about six months to allow the short-lived fission products to decay before the element is sent for reprocessing.

As can be imagined, the design and construction of a charge/discharge machine is exceedingly sophisticated. Not only does it have to be heavily shielded and a good gas-seal applied between the machine and the reactor face but the spent fuel element has to be cooled while it is contained in the machine. The reward for such a complex and expensive machine is the saving in reactor downtime for refuelling.

Altogether eight Magnox stations, each with two identical reactors, were built in England and one in Scotland. They were built by a number of different firms and so the design differs in detail among the stations. The electrical power generated increased considerably with the later reactors and, because of the resulting increase in reactor size, the later reactors to be built, at

Oldbury and Wylfa, were enclosed in a pre-stressed-concrete pressure vessel rather than a welded-steel one. The other main difference in the later stations is that the Wylfa reactor has a gas-cooled spent-fuel storage system rather than a cooling pond.

One drawback to gas-cooled designs is that gas is a rather less efficient coolant than liquid and this imposes an additional constraint on the operating temperature of the reactor. The coolant temperature in a Magnox reactor is about 250°C and the thermal efficiency is thus rather low, about 33 per cent. Magnox reactors have been operating for a considerable time, the original Calder Hall and Chapel Cross reactors (designed mainly as plutonium producers but nevertheless feeding electricity to the national grid) are still operating thirty years after their startup.

The Advanced Gas-cooled Reactors (AGRs)

Almost as soon as the building of the Magnox stations got under way the design of their successors, the AGRs, began. The object was to double the power output of a single reactor from less than 300 Mw(e) to more than 600 Mw(e) in a smaller core, by increasing the operating temperature and also to reduce the amount of reprocessing of fuel by increasing the percentage of the uranium 235 which was "burned up" between fuel changes. These objectives were achieved by using uranium oxide fuel instead of natural uranium metal and canning it in stainless steel. Stainless steel has nearly all the attributes of an ideal canning material except that it soaks up neutrons. The AGR fuel therefore has to be enriched to about 2 per cent uranium 235 content. The reactor is graphite moderated and cooled with carbon dioxide at 40 atmospheres pressure. The outlet temperature of the gas is about 630°C. The fuel content is only about 110 tonnes as compared with more than 300 for the Magnox reactors, nevertheless the reactor has more than twice the electrical output (621Mw) and achieves five times the fuel burnup. Although refuelling is designed to be on load the Nuclear Installations Inspectorate has always refused to permit the AGRs to be refuelled at more than 40 per cent of full power and, on present evidence it is unlikely that full-load refuelling will ever be sanctioned.

The reactor is contained in a pre-stressed-concrete pressure vessel five metres thick. On paper the AGR seemed an almost ideal reactor with a thermal efficiency (42 per cent) comparable to that of conventional stations. In practice the story is rather different. The reactor has been bedevilled throughout its history by cost overruns and long delays in construction. In modern democracies where planning procedures are stringent it takes, at the best of times, about 11 years between the decision to construct a nuclear generating station and the first production of electricity. The first AGR station, Dungeness B, began producing power in 1982, 17 years after it was ordered.

Gas-cooled graphite-moderated reactors were also chosen originally by France which built nine of them. Italy and Japan bought Magnox stations from Britain but by 1970 all three countries had switched to light-water reactors.

Heavy-water-moderated-and-cooled Reactors

The CANDU Reactor

During World War II Canada made a considerable contribution to the Allied atomic research programme and was also engaged in the production of heavy water. After the war she made a policy decision not to become involved in the production of weapons. When the decision was made to proceed with a nuclear power programme the obvious choice of system was a natural uranium (of which Canada had plentiful supplies) heavy-water-moderated system. In the 1950s design and development was begun on a series of reactors which became known as CANDU (CANadian Deuterium Uranium) reactors, and in 1971 the first of four 500 Mw(e) reactors started up at Pickering about 30 kilometres east of Toronto. The size of the Pickering station was later doubled so that the total output is 4000 Mw(e). Another CANDU station on the Bruce Peninsula near Kincardine on Lake Huron has eight 750 Mw(e) reactors.

Other CANDU reactors are operating in New Brunswick, Quebec, Pakistan, South Korea, Argentina and (under construction) in Rumania.

The CANDU reactor is both moderated and cooled by heavy water. The moderator is contained in a cylindrical stainless steel tank (the calandria) lying on its side. Below the calandria is an empty tank into which the moderator can be dumped in an emergency. The heavy-water moderator in the calandria is at near-atmospheric pressure. The calandria is pierced by horizontal tubes going from end to end and these contain both the fuel elements and the cooling water which is pressurised to 85 atmospheres. The fuel is natural uranium oxide canned in zirconium alloy.

CANDU reactors refuel continuously on load. There is a refuelling machine on each end of the calandria and one pushes in a new fuel element while the other receives the displaced spent element. This is eventually taken to permanent storage under water. A safety feature peculiar to the CANDU reactors is the presence of a large building next to the reactor which is not only empty but is leak-tight and kept under continual vacuum. It is large enough to accommodate the vapourised heavy water and volatile fission products which might be released in a severe reactor accident. The same function is achieved by other means in single-unit stations.

CANDU reactors have an enviable safety and reliability record and have for many years headed the "realiability league table" published by the IAEA. In 1984, for example, among all large power reactors in the world, five Candu reactors were listed in the first ten in terms of lifetime performance. Two were ranked first and second. CANDU reactors which have been operating for a year or more operated at 76 per cent of capacity while the world average for other reactors was 58 per cent. In recent years they have been plagued by a series of minor but expensive "plumbing" failures which have necessitated lengthy shut-downs.

Fast Reactors

The fast reactor does not require a moderator but does need a fuel highly enriched (to about 25 per cent) in uranium 235 or plutonium 239. France and Britain have both developed power producing fast reactors. There are numerous advantages — the reactor is compact and will produce several hundred megawatts of electricity from a core the size of a kitchen stove. Removing heat produced at this rate from a small volume requires the use of liquid metals as coolant and sodium or a sodium potassium alloy are currently used and are circulated through the core by electromagnetic pumps which have no moving parts — the liquid metal itself is the only thing which moves. These liquid metals are particularly difficult chemically since they react explosively with water. Any accidental leaks of coolant will ignite spontaneously and, perhaps, explosively with the moisture in the air, and, of course, water cannot be used in dealing with accidental fires. Liquid metal fires are "smothered" with fine sodium chloride particles (surface treated to make them flow easily).

An additional advantage of the fast reactor is the possibility of "breeding" more fissile material. If the core is surrounded by a "blanket" of uranium 238 (otherwise pretty useless as far as generating energy is concerned), some of the fast neutrons produced in fission, which would otherwise have escaped from the core and been absorbed in the biological shield, will be absorbed by uranium 238 atoms and eventually become atoms of plutonium 239 (Figure 13). If the plutonium is chemically separated from the uranium 238 blanket it can itself be used as fuel in other fast reactors. There is even the possibility of a fast-breeder reactor, producing more fuel than it consumes.

In spite of the formidable technological difficulties involved, fast reactor development has been relatively trouble free. The attraction of the system is that it has the potential of increasing the utilisation of uranium resources by more than 50-fold. The present reality is that fast reactor capital costs are much higher than those for thermal stations, the fuel cycle is more expensive because of the cost of reprocessing fuel and it will probably be about 20 years before fast reactors generate electricity as cheaply as thermal systems. Nevertheless intensive development is going on, especially in Britain, France and Russia. In Britain the Dounreay Fast reactor operated for 17 years as an experimental facility with an output of 60Mw(heat) and 25 Mw(e). It was succeeded by the PFR (Prototype Fast Reactor) and the CFR (Commercial Fast Reactor). In France the Rhapsodie fast reactor started operating in January 1967 and was finally decommissioned in 1982. In 1974 the Phénix fast reactor started up at Marcoule and in 1984 the Super Phénix, a 1200 Mw(e) reactor started up at Creys Malville. In the USSR a considerable amount of effort has been devoted to fast-breeder reactors. After a number of experimental fast reactors had been built a power plant, BN-350 (350 Mw(e)) started up at Shevchenko on the Caspian Sea in 1963. Its successor a 600 Mw(e) reactor, BN-600, was constructed at Beloyarsk in the Ural mountains and reached full power in 1981. There are plans for the construction of 800 and 1600 Mw stations.

World Nuclear Electricity Production

Since the Three Mile Island accident of 1979 there has been a hiatus in the hitherto uninterrupted expansion of nuclear energy generation. At that time some plants in the USA were shut down until safety questions were answered and construction of others was postponed. A huge development in Washington was stopped because of cost overruns and orders for new stations in the USA were not forthcoming. In spite of this worldwide nuclear programmes expanded remarkably in 1984 and total installed capacity increased by 17 per cent, the largest annual increase in both relative and absolute terms since the beginning of the large-scale introduction of nuclear power in the late 1960s and early 1970s. During 1984 nuclear power plants produced about 13 per cent of the world's electricity, around 1175 TWh* or about the amount of electricity used by the European Economic Community. There are sixteen countries whose nuclear-generated electricity is in excess of 10 per cent. Thirty-

*TWh = terawatthour = 10^{12} watt hours or 10^6 megawatt hours
GW = gigawatt = 10^9 watts

three countries have nuclear power reactors in operation or under construction and 377 reactors are in operation with an installed capacity of 220 GWe. A further 143 units are under construction with a capacity of 163 GWe. A summary of the percentage of nuclear energy electrical generation is given in Table 7.

Table 7
Commercial Reactors Operating and Under Construction

Country	Reactors Operating	Reactors Under Construction	Nuclear Electricity As % of Total Installed Capacity
Argentina	2	1	13
Belgium	8	—	60
Brazil	1	2	2
Bulgaria	4	2	32
Britain	37	2	18
Canada	15	13	8
China	—	2	—
Czechoslovakia	5	11	15
Finland	4	—	38
France	44	17	65
FR Germany	20	5	31
German DR	5	6	12
Hungary	2	2	5
India	6	2	4
Italy	3	2	4
Japan	32	10	26
Netherlands	1	—	5
Pakistan	—	1	—
Poland	—	2	2
Rumania	—	3	—
South Africa	—	2	4
South Korea	4	5	18
Spain	8	2	22
Sweden	12	—	50
Switzerland	5	—	40
Taiwan	6	—	52
United States	101	17	29
USSR	51	34	11
Yugoslavia	1	—	5

Chapter Eight

Accidents Will Happen

On 28 March 1979 the author answered his telephone and found himself talking to a reporter from a Toronto newspaper. The conversation went like this:

> "How many deaths will be caused in Ontario by the Three Mile Island accident?"
>
> "I don't know anything about the accident except what I have heard on the radio, but I don't believe that any deaths will be caused in Ontario."
>
> "Oh, well how many deaths will be caused in Pennsylvania?"
>
> "I don't think that any deaths will occur in Pennsylvania either."
>
> "Do you mean that there isn't a story then?"
>
> "Well, unless they are very lucky they will have written off the reactor."

The reporter cheered up a bit at that and reported the last sentence in a story that day. In the end, the opinions expressed proved to be an accurate prediction. The conversation is unimportant except that it reflects a media attitude that is rather prevalent. It is news of disasters that sells newspapers and makes news, and the nearer the disaster is to home the better. The media must bear some of the responsibility for the somewhat irrational attitude of the public to nuclear energy.

There has perhaps been no industry in history which was set up with such an appreciation of the risks involved or with such a large effort mounted to obviate them as the nuclear industry. Nevertheless accidents have occurred and will continue to occur. Every accident must be studied meticulously to determine what went wrong and measures taken to ensure that the same sequence of events is prevented in the future, either by better design or better operating practice. There is nothing new in this. The same procedures have been used for more than a hundred and fifty years on the railways and for more than fifty years in commercial airline operation and in dam construction. Failure of equipment or human error in any of these endeavours can result in deaths and injury on a scale comparable to the worst that could be expected from an accident to a reactor. However human beings are not rational in their assessment of personal risk and the prospect of being exposed to radiation from an accident to a reactor carries an extra dimension of fear of the unseen and the unknown which is not often encountered in railway stations or airports.

The first defence against a possible accident is the integrity of the design of the reactor. It is an *a priori* assumption that reactor design engineers will design for safety, will anticipate failures of equipment and accidental and even deliberate maloperations and build in safety devices to prevent them. All safety circuits will be at least duplicated and provided with an independent emergency power supply. Safety devices and interlocks must be well maintained and regularly tested. There have been many instances of emergency equipment failing to start when required simply because it is so very rarely needed that it has never been switched on. In addition it is imperative that the design and construction of the reactor will be overseen by an independent regulatory body which will carry out a continuing independent safety assessment before granting a licence for the construction and operation of the reactor. In Britain this function is discharged by the Nuclear Installations Inspectorate, in the USA by the Nuclear Regulatory Commission and in Canada by the Atomic Energy Control Board.

The First Nuclear Accident Victim

On the morning of Friday, 13 July 1945, perhaps the most talented and certainly the tensest group of scientists ever assembled was gathered in the New Mexican desert to watch the assembly of the core of the first atomic bomb. At 9 am the final assembly began. The team of eight scientists hovered round the table where the pieces of plutonium were laid out. Louis Slotin's job was to put the pieces of the bomb's core together so that they formed a near-critical mass.

The atmosphere in the room was electric. The successful culmination of the most extraordinary scientific and engineering project in history depended on the knowledge and skill of this 35-year-old Canadian. Should he make a slip the test would be doomed, but so would the watching scientists, from radiation sickness. Their presence was an entirely unnecessary and irresponsible risk. It seems highly reprehensible now, and especially in the light of what happened a few months later, that they should have been allowed anywhere near the assembly. Even Oppenheimer came to watch, but now everything was out of his hands and he was so fidgety that he was asked to leave. The assembly was completed without incident and the bomb successfully tested.

Louis Slotin went back to Los Alamos where he had made a specialty of the tricky and dangerous work of making experimental critical assemblies. It was a nerve-racking business which he seemed to enjoy. Other scientists couldn't stand it and Fermi would take his group on a camping trip whenever a critical assembly experiment was planned. On 21 May 1946 Slotin was preparing to leave Los Alamos for he had just been appointed an Assistant Professor at the University of Chicago. He was demonstrating an experimental assembly to Alvin Graves who was going to take over from him when he left. The task was to bring two sub-critical hemispheres of plutonium almost together so that a very slow and controlled chain reaction would take place. Incredible, as it seems now, this was done by hand and Slotin was going to lower one of the sub-critical hemispheres until it almost touched the other. The two were prevented from actually touching by two metal spacers on the lower hemisphere. For some reason Slotin removed the metal spacers and substituted the blade of a screwdriver which he held in his right hand, while he lowered the upper hemisphere with his left. As they approached criticality Slotin edged out the screwdriver gently by wiggling it from side to side, so the two hemispheres came closer and closer.

Suddenly it happened. The screwdriver slipped out and the two hemispheres came into contact. A blue glow filled the laboratory and the neutron monitor went mad. A searing flash of heat swept across the lab. Slotin knew what had happened and instantly knocked the plutonium pieces apart with his bare hands. The chain reaction ended and everything seemed as before. Slotin knew that nothing would ever be the same again. He called for an ambulance and, within minutes he and Graves were in the Los Alamos Hospital. Slotin quickly became nauseous, but after the first day his general condition seemed to improve. He told Graves that he was very sorry to have got him into this mess and, while his own chances of survival were only 50/50, he hoped that Graves' were better than that.

His hands and arms had received the greatest dose and swelled up rapidly. After five days his white cell blood count had fallen disastrously in spite of daily transfusions and it was realised that he would die. After the seventh day he became disoriented and confused and on 30 May 1946, the ninth day after his irradiation, he died. It was established from the radioactivity induced in a ring he wore, in coins in his pockets and from the autopsy, that he had incurred a dose of about 900 rem.

Alvin Graves had received a dose of about 200 rem. In hospital he vomited once but his nausea cleared up after about eight hours. After five days his temperature rose to 103°F and he had two blood transfusions. He suffered from insomnia and was very irritable but he recovered gradually and was able to go home after ten days. He had some hair loss and was sterile for a time but he made a nearly complete recovery and returned to scientific work.

Thus ended the first nuclear accident. Like many others to follow it was caused by overconfidence, complacency and the circumventing of safety devices. We can learn from it still.

The SL1 Accident

Until Chernobyl the only reactor accident which caused immediate casualties was that to the SL1 reactor at the US National Reactor Testing Station north of Idaho Falls in January

1961. The SL1 was a 3Mw(t) boiling-water military reactor which was a prototype for a small reactor designed to provide electricity and space heating in remote locations. On 3 January 1961 it had been shut down for maintenance since 23 December 1960. The work had been completed and it was planned that the reactor would again be brought to power on 4 January. The crew of the 4 pm to midnight shift on 3 January consisted of three military personnel, John Byrnes, Richard McKinley and Richard Legg, one of them a qualified chief operator, another a trained operator and the third a trainee. They were assigned the task of reassembling the control rod drives and preparing the reactor for startup. The first indication of trouble was an automatic alarm received at the security headquarters at 9.01 pm.

At the same time the personnel radiation monitor at the Gas-cooled Reactor Experiment about one mile away from the SL1 reactor alarmed. Upon entering the SL1 fenced area the fire crew were unable to alert the SL1 operators. The assistant fire chief entered the support building and immediately detected radiation levels of 25 rem per hour. The health physicist then proceeded towards the reactor building and encountered a radiation field of 200 rem per hour at the staircase to the reactor. It was decided to enter the building to locate the operators and two of the crew were found on the floor near the reactor in a radiation field of approximately 1000 rem per hour. Richard Legg was dead but John Byrnes was still alive and was brought out of the reactor building by about 11 pm. He died shortly afterwards from the effects of a head wound. McKinley's body could not at first be located but was later discovered pinned to the reactor building roof by an ejected control rod plug. The results of post mortems on the three bodies showed that McKinley and Legg died instantly from the results of an explosion and that Byrnes' head wound precluded any possibility of survival. During the rescue operations and subsequent entries to recover equipment and records 22 out of several hundred people involved incurred radiation exposures of from 3 to 27 rems.

The cause of the SL1 accident is still not completely clear. It is known that the tasks assigned the operators involved lifting the control rods. The chief operator and the operator had performed this same task at least four times before and had been specifically trained for it. Their training included the instruction that the control rod was not to be raised more than four inches. It would have been necessary to lift the control rod 26 inches for the reactor to go critical and for another 6 to 8 inches at a rate of about 24 inches per second to produce a nuclear excursion of the magnitude which was estimated to have occurred. From the positions of the men after the incident and from the injuries they received it is possible that Legg and McKinley were actually lifting the control rod at the time of the explosion. It may even have happened that the control rod stuck and the men were trying to free it by pulling as hard as they could when it suddenly came free. There was some evidence that the control rods of this reactor had previously tended to stick on occasion.

The reactors at NRTS are spread over an area of about 1000 square miles and are quite remote from one another; to visit one building from another often entails a 30-minute drive. There are no local inhabitants, apart from reactor operating personnel. Measurements made after the accident suggested that a narrow plume of gaseous fission products travelled south-southwest from the reactor building and low-level activity of sage brush due to iodine 131 was observed. Soil samples indicated low contamination by strontium 90 for some time after the accident.

The Windscale Accident

By 1957, twelve years after the explosive dawn of the nuclear age, it seemed that mankind was well on the way to the unimagined riches of unlimited and cheap electrical energy. The first United Nations Conference on the Peaceful Uses of Atomic Energy had been held in Geneva in 1955 and there was an almost universal feeling that from now on governments would cooperate in the solution of the world's energy and food problems. There was also a sense that technology could conquer everything, just as it had tamed the atom. Then came the Windscale Incident (it was some years before it was referred to as an accident) and the sudden realisation that perhaps the atom had not really been tamed after all. Most people still connected a nuclear pile with the atomic bomb — a sort of atomic bomb in a strong container, which generated energy without causing pollution. Gone were the days of power stations churning out black plumes of smoke. If people thought about the dangers of nuclear energy, they did so in terms of a possible explosion rather than a release of radioactive material to the environment. In October 1957 there was a sad awakening when the realisation came that sometimes nuclear reactors could pollute and that the pollution was in some ways more insidious than the old black smoke just because one couldn't see it or smell it or even know that it was there.

The Windscale Reactors

The first British production reactors were located at Windscale on the Cumberland coast about 20 miles south of Whitehaven. There were two graphite-moderated air-cooled "straight through" reactors (i.e., the air coolant was not recirculated but was blown straight out to the atmosphere). When the reactors were designed it was realised that any airborne particulates sucked into the reactor would be irradiated and made radioactive before being released to the atmosphere. Inlet filters were therefore installed to remove the atmospheric particulate. It was not originally intended to filter the air again before release, but rumor has it that Sir John Cockcroft on a visit to the US establishment at Hanford saw a large hole being dug beside the site of a similar reactor. When he enquired what the purpose of this was he was told it was to accommodate the exhaust filters. He is then reputed to have made a surreptitious lunchtime visit to the Richland Post Office to send a cable to Sir Christopher Hinton (who was in charge of the construction of the reactors) saying "put exhaust filters on the Windscale reactors." By this time the reactor stacks were already 200 feet high and the only place left to put the filters was at the top, 400 feet above ground level. It was then necessary to tack an elevator and a staircase on to the stack so that the filters could be serviced. They and the filter gallery can be seen clearly in Figure 16.

Whatever the truth of this story it is certainly plausible. It is

Figure 16 The Windscale Piles

not easy to think of another reason why a filter gallery should be sited at the top of a stack where the available filter area is so limited that the air has to be filtered at an inconveniently high velocity. The filter units were in two foot square frames located in 40 rows (angled at about 30° to the vertical to increase the effective filter area) across the stack and they were changed continuously by using a hydraulic ram to push in a clean element. A dirty element was pushed out the other side of the gallery into a water trough where it was picked up by a travelling arm and brought to the end of the trough to be hosed through. By 1951 the material used in the original filters was no longer available and the author's first scientific task on arrival at Windscale as a newly qualified physicist was to design new filters for the Windscale stacks.

Wigner Stored Energy

The Windscale reactors were designed as plutonium producers, the fuel was natural uranium canned in aluminum alloy, and the operating temperature was rather low by present day standards. When graphite slows down neutrons some of the energy of the neutrons appears directly as heat and some is used to knock graphite atoms from their normal positions in the lattice into other positions of higher potential energy. Thus some of the energy of the neutrons is stored in the graphite. This phenomenon was not known when the Windscale reactors were designed although it was predicted soon afterwards on theoretical grounds by Wigner, and was known thereafter as "Wigner energy." As the reactor continues to operate more and more energy is stored in the graphite and, eventually, if nothing is done, the displaced atoms will return to their original sites and the stored energy will be released as heat. This could be a serious embarrassment, if it happened unexpectedly or in an uncontrolled manner, for it might cause the graphite temperatures to rise to unacceptable levels and even trigger the melting of some fuel cans.

A spontaneous release of Wigner energy had, in fact, occurred in Pile 1 in September 1952 while the reactor was shut down. This led to a rise in the temperature of the graphite but the rise was not dangerous and no harmful effects resulted. As a result of the study of this incident a programme was instituted for periodic controlled releases on each of the reactors. The procedure was fairly straightforward. The Wigner energy release could be triggered by raising the graphite operating temperature above normal. To follow the progress of the energy release it was necessary to have more complete information on its temperature distribution than was needed for the normal operation of the reactor. The reactor was therefore shut down and extra thermocouples installed in the moderator at carefully calculated positions. The reactor was then started up again with no coolant air flow thus raising the uranium and graphite temperatures to trigger the Wigner release which was self sustaining once started. The reactor was then shut down and the Wigner release proceeded. It was a delicately balanced operation, for, because of the time lag in the temperature response of the graphite, its temperature would continue to rise after the reactor was shut down. If the reactor were shut down too soon the graphite temperature would not rise sufficiently to anneal the whole moderator; if too much nuclear heating was applied there was a danger that some fuel elements would overheat and the cans might leak. Nevertheless, by October 1957 the procedure had been performed more than a dozen times and was regarded as fairly routine.

The Release of October 1957

A Wigner energy release was started on Windscale Pile 1 late on 7 October 1957. By 11 am on 8 October the engineer in charge of the operation thought that the release was dying away and that parts of the graphite moderator were going to escape being annealed. What he did not know was that the thermocouples were not installed in the hottest part of the core. As the subsequent enquiry showed, a second nuclear heating was applied at too rapid a rate and probably caused the failure of one or more fuel cans, whose contents quietly oxidised away. There was no indication on the reactor instruments that this had taken place until the afternoon of Thursday 10 October when the stack radioactivity monitor alarmed and the staff thought they had a bad can burst on their hands. It was much worse than that. At 4.30 pm visual inspection through a plug hole in the pile revealed glowing fuel elements and, through another viewing port, flames could be seen shooting out of the discharge face of the reactor. In desperation every method that could be thought of was tried to extinguish the fire — shut off the air, blow it out with carbon dioxide (a tanker of liquid carbon dioxide had just arrived on site to top up the coolant in the Calder Hall reactors next door). Nothing worked.

By midnight the major decision had been made that, if all else failed, water would be used to try to extinguish the fire. Fire hoses were dragged up the charge face of the reactor and coupled to the entry ports of a line of fuel channels about a metre above the heart of the fire. The water was turned on at 8.55 am on Friday 11 October and off at 3.10 pm on Saturday 12, by which time the pile was cold. By this time 27,000 curies of iodine 131, 12,000 curies of tellurium 132 (which decays with a half life of 78 hours to its daughter iodine 132, itself radioactive with a half life of 2.3 hours), 600 curies of caesium 137, 200 curies of strontium 89 and 2 curies of strontium 90 had been released to the atmosphere and was being carried away by the wind and deposited on the ground. (Chamberlain, 1981)

The Effects

From midday on 10 October the winds at Windscale were light and variable with a tendency to blow from the southwest. As the accident proceeded the weather changed and the winds quickly veered to a northerly direction and gradually backed to the northwest during the morning of 11 October. Some of the radioactive material released early in the accident was thus carried northeast (Figure 17) producing a zone of deposition as far east as Yorkshire. Material released later in the incident was blown to the southeast causing the main deposition of radioactive iodine over Cumberland and Lancashire. The radioactive material (by then at low concentration) reached Belgium late on 11 October. Frankfurt late on 12 October and eventually circled back to Norway at some time on 15 October (Figure 18).

The immediate problems facing the health physicists were to control the dose received by the workers on the reactor and to take measures to protect the health of the public in the neighbourhood of the station. Men working on the reactor were in protective clothing and wore respirators and their periods of exposure were strictly controlled. As a result measurement of the thyroid doses received by 96 men showed that the maximum

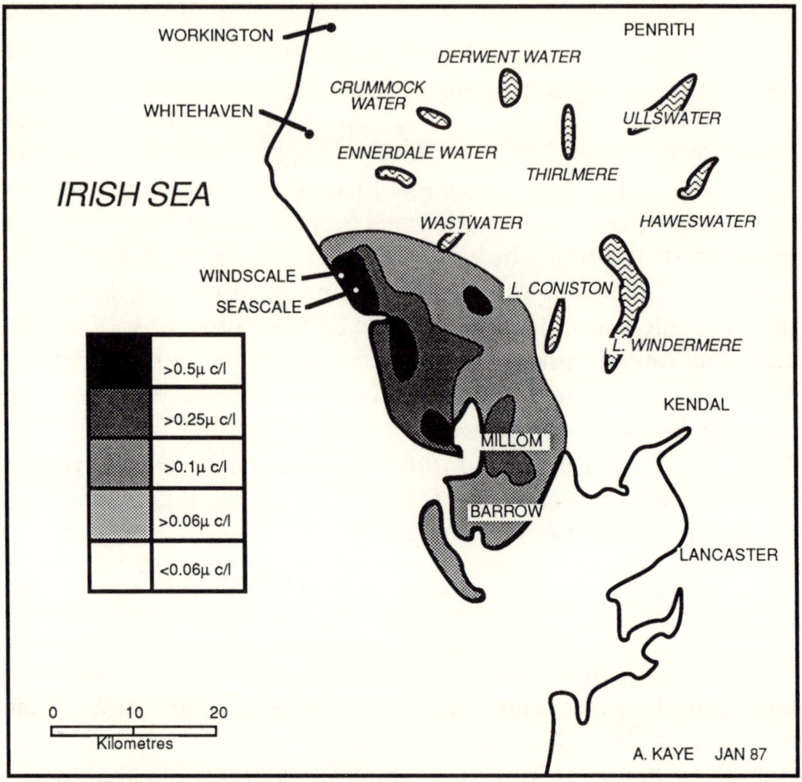

Figure 17 Radioiodine Contamination of Milk Around Windscale

thyroid dose received was 9.5 rem, the next highest 2.1 and the average 0.4 rem. The external radiation dose control was such that during the 13-week period including the accident, only 14 workers exceeded the maximum permissible dose of 3 rem and the highest dose was 4.7 rems.

Measurement of the direct radiation dose under the plume showed that there was no need to evacuate the population downwind of the reactor. Later calculations showed that the integrated dose to people living in the area of maximum deposition was about 50 millirem (as compared to the natural dose of about 110 millirem per year from cosmic rays, radium in building materials and radon emitted from the ground). There was also a small additional external dose from the passing cloud.

It was apparent that the major hazard would be to small children drinking milk from cows grazing on grass on which iodine 131 had been deposited and it was decided that the distribution of milk from an area of about 200 square miles downwind of the plant should be controlled. In the event the milk was collected as normal and poured down the drain. It

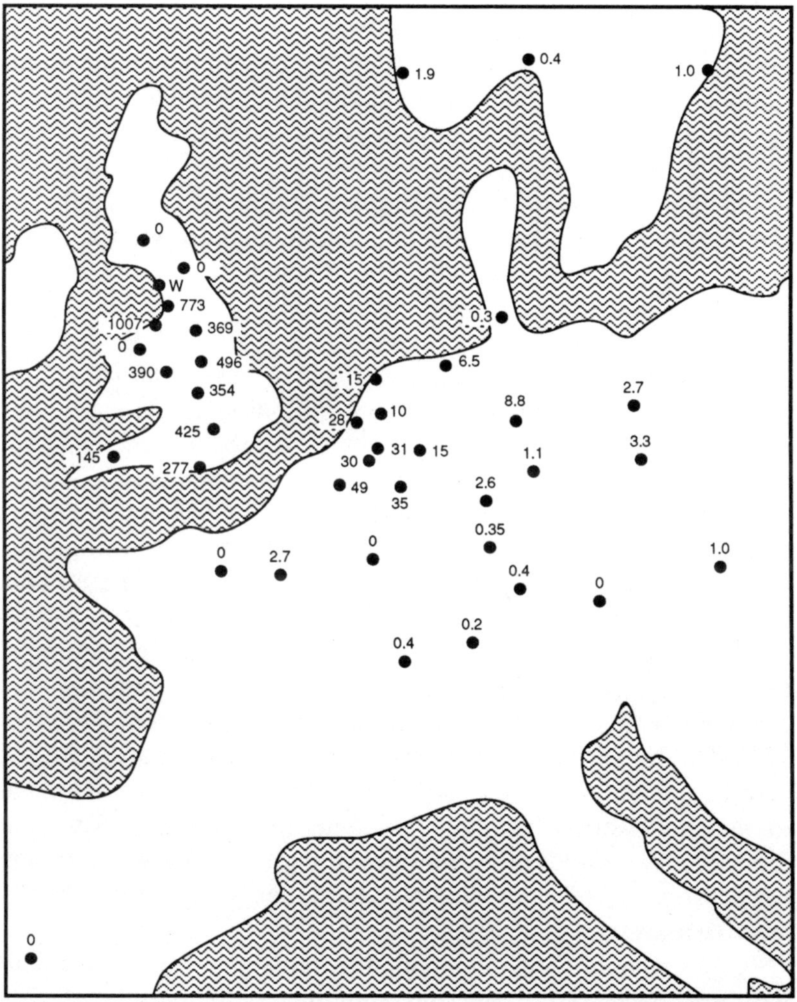

Figure 18 The Spread of Windscale Radioiodine Around Europe. (The figures give the time integral of iodine concentration in air in micro-microcurie-days per cubic metre).

would have been medically acceptable for the milk to be processed into dried milk, cheese or chocolate and the products stored for three months until the iodine 131 had decayed. The Ministry of Agriculture apparently decided that the public relations problems of distributing products which had been contaminated with iodine 131, even if it had all decayed away, were too great to make the exercise worthwile.

In all, more than two million litres were poured into drains. The farmers were compensated by the government and it was rumoured that once the farmers realised what was happening to their milk, the output of milk from the cows in the affected area increased dramatically.

The highest level of iodine recorded in milk was 52,000 Bq per litre and the distribution to the public was prohibited at levels greater than 3700 Bq per litre. The Ministry of Agriculture permitted the resumption of milk distribution when all the samples of milk from a given area contained less than 3700 Bq per litre and the activity of the milk was decreasing with a half life of not more than eight days. The half life of iodine 131 is eight days and this second restriction was applied to preclude the distribution of any milk containing longer lived isotopes such as strontium or caesium. The contamination had dropped to these levels throughout most of the restricted area after 25 days but the control was not removed from the most highly contaminated area for 44 days which allowed time to carry out sufficient analyses for strontium to be sure that restrictions need not be continued because of strontium 90. Samples of drinking water were also analysed from reservoirs and domestic supplies. The levels of radioactivity varied from the limit of detection (about 4 Bq per litre) to around 40 Bq per litre. The potential hazard to members of the public was thus very small compared to that from contaminated milk. Samples of eggs, vegetables and meat were also examined and shown to be rather insignificant sources relative to milk.

In some cases there was reason to suspect that the controls had not been fully effective and about 200 adults and 30 children were invited to have dose determinations done on their thyroids. The highest dose to a childs thyroid was estimated at 16 rem and to an adult thyroid at 4 rem.

Conclusions

That the major hazard from Windscale was due to iodine 131 in milk is due to the fact that, because of its volatility, iodine is preferentially released from irradiated uranium at high temperatures. It was first reported (Beattie, 1959) that the stack filters retained between 20,000 and 30,000 curies of iodine 131. A later review of various analyses of samples of the filters which has been made by Chamberlain (1981) suggests that only about 2,600 curies of iodine 131 was retained by the filters. Patterson's (1983) statement "As it turned out 'Cockcroft's Folly' (i.e., the stack filters) kept a major accident from becoming a catastrophe" appears to be incorrect. Chamberlain concluded that about 27,000 curies of iodine 131 were released together with other isotopes which were volatile at temperatures up to 1000°C or which had volatile oxides. He considered that the stack filters had similar efficiencies for isotopes of iodine, caesium and polonium since the ratios of these elements on the filters and in environmental samples were similar and that the efficiency of the filters was between 15 per cent and 50 per cent. The total release of strontium was about 200 curies (0.1 per cent of that in the affected channels), a considerable increase on the 80-curie figure reported earlier by Loutit et al (1960). A few hundred curies of polonium 210 were released to the atmosphere, probably in the form of sub-micrometre particles. Very little uranium oxide particulate was released, or even reached the filters. It was estimated that iodine 131 was released about a hundred times more readily than the strontium isotopes. It was found from experimental studies that 1000 Bq of iodine 131 deposited per square metre of grass would lead three days later to a concentration of 170 Bq per litre of milk. Weather conditions and feeding patterns subsequent to an accident can alter this figure. In the first two weeks after the Windscale accident, for example, rain reduced the quantity of iodine 131 remaining on grass by 50 per cent. The proportion of a cow's feed derived from pasture also bears a direct relation to the activity in milk.

In many ways and for a number of years it seemed that, if you had to have a reactor accident, one like Windscale was almost ideal. There were no apparent casualties and, while the accident was very costly in monetary terms (two valuable reactors were written off, for it was decided that it was too expensive to modify Pile 2 to the required safety standards and both reactors were therefore sealed off), the gain in knowledge was enormous. There was also a sudden healthy upsurge in interest in reactor safety, for, after a dozen years of practically trouble free reactor operations, many reactor operators were becoming a little complacent.

Long-term Effects

In 1982, with the benefit of 25 years of further research into the effects of the low-level radiation on people, the UK National Radiological Protection Board (NRPB) made a further study of the public health consequences of the accident. Crick and Linsley (1982) concluded that the population of Britain as a whole received an extra dose of radiation to the thyroid equivalent to about half the dose it would normally receive in a year from natural radiation. They considered, on the basis of pessimistic assumptions, that the iodine released could have caused the number of cases of thyroid cancer to increase by as much as one per cent. Most cases of thyroid cancer are cured and the number of deaths from thyroid cancer occurring every year in Britain is between 15 and 25. The NRPB report estimates that the Windscale accident may have caused an additional cancer death every three years, and that the upper limit on the total eventual additional death from cancer as a result of the accident was 20.

Shortly after this report was published it was revealed that samples of the element bismuth were being irradiated in the Windscale reactors at the time of the accident to make polonium. The release of polonium was not taken into account at the time of Crick and Linsley's report, but an addendum (September 1983) shows that, as a result of the polonium (the dose from which was estimated to be one tenth the annual dose arising from naturally occurring polonium), the theoretical upper limit of numbers of fatal cancers rose from 20 to 33.

The report emphasises that at low doses and dose rates only an upper estimate can be made of health effects in the population. The actual number of health effects is likely to be lower and may be zero. The addendum also shows that over 95 per cent of the collective dose was contributed by exposures equal to those received in a period of less than three weeks from exposure to natural background radiation. If this small additional dose makes any contribution at all to ill health it will be spread over several decades and will amount to less than one tenth of one per cent of the health effects attributable to the natural background radiation on the same theoretical basis.

Three Mile Island

At approximately 4 am on Wednesday 28 March 1979 an accident involving reactor unit 2 occurred at the Three Mile Island Nuclear Generating Station near Harrisburg, Pennsylvania (Figure 19). The public health consequences of the accident were negligible but the public relations aspects of it were so badly

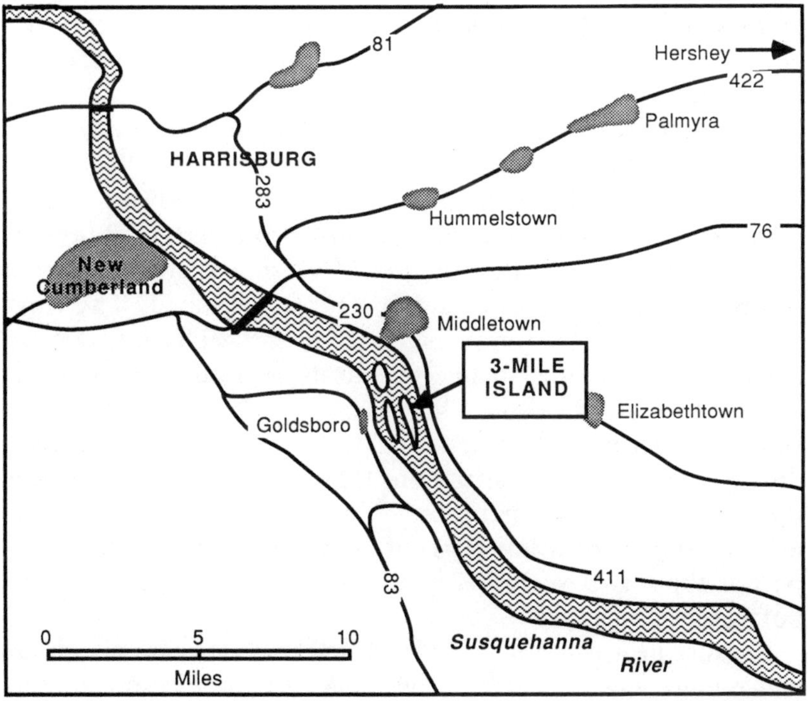

Figure 19 The Three Mile Island Site

handled and the subsequent enquiry revealed such operator unpreparedness that the accident had a major impact on public acceptance of nuclear energy and indeed on energy policy in the USA. For these reasons alone it merits close study.

The Reactor

The reactor design is shown in Figure 20. It is a fairly standard PWR in which the heat generated by the uranium oxide is carried away by pressurised water in the primary loop and passed through two steam generators where steam is raised in a separate secondary circuit to feed the steam turbines. The water pressure in the primary loop RCS (reactor coolant system) is controlled at around 150 atmospheres in order to maintain the water as liquid at its temperature of 315°C. The primary circuit is fitted with a pressuriser to control its pressure and the water is circulated by four reactor coolant pumps. The built-in safety features include a fast automatic shut-down system which quickly reduces the heat generated by the reactor fuel to a few per cent of the normal level and both high- and low-pressure emergency core-cooling systems (ECCS) which automatically inject water into the primary cooling circuit to replace any water which may have escaped.

At 4 am on 28 March maintenance work on feedwater equipment to the secondary circuit caused an inadvertent shutdown of both main feedwater pumps. The steam generator thus stopped producing steam and the turbine tripped. The three auxiliary feed water pumps started automatically but did not deliver water to the steam generator because two manually operated isolating valves, which were in an accessible area outside the reactor containment building, were not open. According to the operating instructions these valves should have been open during reactor operation. It was reported that they were left closed in error following routine testing of the auxiliary feed pumps about two weeks before the accident. The operators were not aware of this.

Because heat was not being extracted from the primary circuit (since there was no water in the secondary circuit) the pressure in it rose until an electromagnetic relief valve on the pressuriser

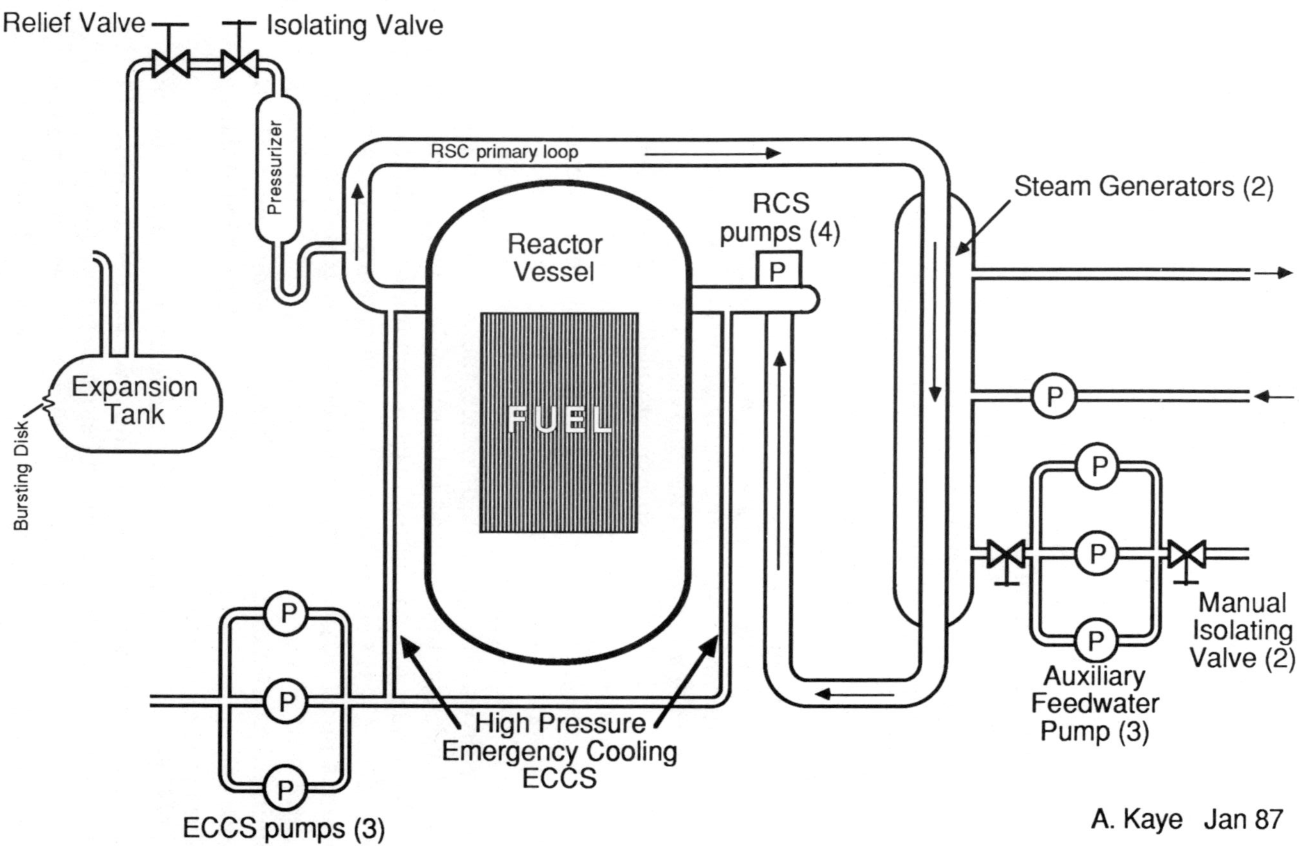

Figure 20 The Three Mile Island Reactor

opened and the reactor shut down automatically. Primary coolant water passed through the relief valve to a closed expansion tank and the primary circuit pressure dropped. If the relief valve had then reclosed as it was designed to do and if feed water had been delivered to the steam generator by one or more of the auxiliary pumps, the incident would have ended at this point.

The relief valve stuck open and the primary circuit pressure continued to drop until at least two of the emergency cooling system high-pressure pumps began injecting water into the primary circuit. The quantity of water delivered by the pumps was greater than that escaping through the open valve and this caused the indicated water level in the pressuriser to rise until the operators were concerned about loss of pressure control.

At about 4.05 am the operators shut down the first emergency pump and five minutes later the second one. Pressure in the primary circuit then fell rapidly. At about 4.08 am the operators restored feedwater flow to the steam generator by opening the manual isolating valves. This lowered the primary circuit temperature and the pressure continued to drop owing to discharge through the open relief valve, even though the high-pressure emergency pumps were turned on again at about 4.12 am. Heat was now being transferred from the reactor fuel.

At about 4.15 am the bursting disc on the expansion tank burst, as it was designed to do when the pressure rose to the design level, and this allowed primary coolant water to pass to an open sump area within the reactor containment building. As the level rose in the sump, the water was automatically transferred to tanks in an auxiliary building outside the reactor containment. Primary coolant flow to the expansion tank and the sump continued until about 6.20 am when the operator isolated the stuck-open relief valve by closing a remotely operated isolating valve in the relief line.

The operators shut down two of the reactor coolant pumps at about 5.15 am and the remaining two at 5.40 because the primary circuit pressure dropped below suction requirements of these pumps. This was an almost fatal mistake because it stopped primary coolant flow through the reactor vessel and it is thought that coolant in the neighbourhood of the fuel probably boiled, reducing heat transfer from the fuel, allowing some fuel cans to rise in temperature and fail. The hot zirconium then produced hydrogen gas in a chemical reaction with water. The hydrogen, together with some radioactive gases released from the damaged fuel formed a "bubble" at the top of the reactor vessel.

At approximately 8 pm the operators established stable reactor cooling conditions with the relief valve isolated, one reactor coolant pump operating for coolant circulation through one steam generator and feed water flow to and steam flow from the steam generator to the main condenser.

The total release of radioactive materials from the station after the accident was very small. Although there were transient rises of the containment building pressure, there was apparently no escape of radioactive materials directly from the containment building to areas outside the station. Most of the externally released radioactive materials came from tanks inside an auxiliary building.

Assessment of Events of 28 March 1979

The accident would have ended with a reactor trip after the main feed water pump failure had it not been for two further equipment failures — the loss of all feed water supply and the stuck-open relief valve.

The important reactor safety systems responded automatically as designed; the reactor tripped promptly, the emergency cooling system replaced liquid lost from the primary circuit and the reactor containment limited releases of radioactivity so that public doses were extremely small.

Early operator reactions were not adequate to prevent significant fuel damage which in turn led to the escape of radioactive water and gases to the auxiliary building tanks, and to the formation of a "bubble" of hydrogen and radioactive gases which accumulated at the top of the pressure vessel. Some of the actions, or absence of actions which contributed to the seriousness of the accident were the shutting down of the emergency coolant pumps; the shutting down of the main coolant pumps; delay in isolating the open relief valve; delay in starting a main coolant pump in order to reestablish reactor cooling.

Assessment of Events After 28 March

After reactor conditions were stabilised on the evening of 28 March the main problem was lack of knowledge of the size and composition of the gas bubble in the reactor vessel. There was concern that sufficient oxygen might enter the bubble to create an explosive hydrogen-oxygen mixture with the potential of causing more fuel damage and, perhaps, breaching the pressure vessel. These fears were triggered by an event at 2 pm on 28 March when hydrogen escaped into the containment through the open relief valve via the open dump tank, mixed with air and caused an explosion which showed up as a 28psi pressure peak on instruments in the control room. This led members of the Nuclear Regulatory Commission (NRC) team to ask what would happen in hydrogen within the reactor were also to explode. It would of course need oxygen to do so but the range of explosive mixtures was so wide that someone in the NRC quickly provided a theory (which later proved to be wrong) that the radiation within the reactor would cause decomposition of water into hydrogen and oxygen and thus provide the ingredients for an explosion which might blow the pressure vessel apart.

On 30 March Governor Thornburgh advised that children and pregnant women living in the vicinity of Three Mile Island should leave and many did so. Most of these made private arrangements and an evacuation centre at Hershey about 15 miles distant only attracted about two hundred clients.

Release of Radioactive Material to the Environment

Most of the airborne radioactive material released came from the waste-storage tanks in the auxiliary building to which it had been transfered from the reactor containment. All the releases had passed through high-efficiency filters which removed particulates and charcoal traps which removed iodine. The airborne releases consisted almost entirely of the radioactive noble gases xenon and krypton, primarily xenon 133. These give rise to external radiation doses and are not absorbed internally by the human body. The Nuclear Regulatory Commission estimated that the total release of xenon 133 was 1.3×10^7 curies. The total release of iodine 131 to the atmosphere during the period 28 March to 7 April was estimated to be about 1.4 curies, although other estimates have been from 15 to 30 curies. This is a very small quantity of iodine compared to the 27,000 curies released at Windscale. Slightly contaminated waste water was discharged from the turbine building into the Susquehanna River. It has been stated that at all times the concentration of radioactivity in the liquids discharged was within the regulatory limits for normal operation.

Radiation Dose to the General Public

Ground level gamma dose rates at the site boundary reached levels of 25 millirem per hour during the morning of 30 March. Levels off-site and in populated areas were typically less than one mrem per hour except on the morning of 30 March when off-site gamma dose rates were reported to be a few millirem per hour. After 1 April levels at off-site locations were typically less than 0.2 millirem per hour. These dose rates were caused by radioactive noble gases in the overhead plume as there was essentially no ground deposition of radioactive materials. Of over 150 air samples taken off-site during the 28 March to 2 April period, 8 showed measurable quantities of iodine 131, ranging from 0.01 to 1 Bq per cubic metre. The maximum permissible concentration of iodine 131 for continuous breathing by members of the public is 3.7 Bq per cubic metre of air.

Milk samples were taken and analysed independently by the Food and Drugs Administration. Most samples showed no detectable contamination and the maximum iodine 131 level found was 1.5 Bq per litre. The US Department of Health, Education and Welfare has set a level of 450 Bq per litre as that for withdrawing milk from market.

None of the more than one hundred water samples taken and analysed by 3 April showed detectable radioiodine. Of the several hundred vegetation and soil samples collected and analysed, only a few grass samples showed the presence of iodine 131 and these were barely over the detectable level.

Dose Estimates

The maximum whole-body dose which could have been received by any member of the public was estimated by the Nuclear Regulatory Commission to be less than a hundred mrem which is about the same dose as a person receives in a year from natural radiation. This estimate assumes that the individual was outdoors for 24 hours per day at the location of greatest exposure. The maximum thyroid dose which could have been received by a young child by inhalation was estimated to be less than 5 mrem and the total dose received by a child drinking the most contaminated milk found was also estimated to be less than 5 mrem.

The Aftermath

Seen in the perspective of time the Three Mile Island accident caused a great deal of anxiety which was unjustified. The emergency was really over by 8 pm on 28 March when conditions in the reactor were stabilised and proper cooling reestablished, but for those closely involved there was nearly a week of uncertainty and worry. The chief cause for worry was the uncertainty as to what was really happening and this feeling was shared even by some who were closely involved. In spite of the fact that no one was injured or is likely to be injured in the future by radiation received from the accident, the nuclear industry in the USA was greatly affected. Some orders for nuclear power stations were cancelled and others postponed.

On the positive side there is scarcely a nuclear reactor anywhere in the world whose safety has not been reviewed since Three Mile Island to determine whether the same thing could happen to it and how well it would respond to such an emergency. There are many things to be learned from the accident. First and foremost, it is the simple things which go wrong, there is no point in having safety equipment if it doesn't work, and one cannot be sure of it working if it is not tested regularly. Secondly, more thought has to be given to the design of control rooms. At Three Mile Island red lights sometimes indicated a satisfactory state of affairs and at other times that something was very wrong. It is standard practice in the engineering industry that an open valve is indicated by a red light. A green light, which to most people says that all is well may mean the opposite when it indicates that a certain valve is closed.

Finally it is not just sufficient to train reactor operators once. Training must be continuous and comprehensive. It must not just involve the reactor operators, but everyone employed on the site so that all employees know their roles in an emergency and are confident of their ability to fulfil them.

Other but Less Catastrophic Accidents

The following is a summary of other known incidents in both research and production reactors which did not lead to serious environmental contamination and of a still mysterious accident to a nuclear facility in the USSR which apparently did.

Canada — Chalk River (experimental), 2 December 1952

An employee mistakenly lifted four of the twelve control rods out of the fuel core of the NRX experimental reactor causing a chain reaction that melted part of the core. A million gallons of radioactive water accumulated inside the reactor building and the cleanup took six months.

Cause: Human error

USA EBR-1 (experimental), November 1955

The EBR-1 experimental enriched uranium fast-breeder reactor at NRTS, Idaho Falls went out of control during tests because of misshapen fuel rods and human error.

Cause: Fuel rods and human error

USSR — unreported, winter 1957–58, Kyshtym, Chelyabinsk

A major nuclear accident in the Ural Mountains. Little information has been released but the accident is thought to have occurred at a nuclear weapons facility and resulted in the contamination of as much as 400 square miles. There are no reliable estimates of casualties but several villages have been abandoned and a river rerouted to avoid contamination.

Cause: It is surmised that radioactive wastes were dumped in abandoned mines in the Ural Mountains and the decay heat generated steam pressure which eventually caused a devastating mud blowout which blanketed the surrounding region with highly radioactive effluent. It is difficult to believe that an accident having such a widespread effect could have happened in this way as nuclear waste does not contain enough energy. The suspicion must remain that the cause was an accident in a defence plant. No admission of this accident has ever been made by the Russian government.

Canada — Chalk River, 23 May 1958

A defective fuel rod over-heated after being removed from the NRU reactor. Some radioactivity released to the environment. A long clean up followed.

Cause: Defective fuel rod over-heated.

USA Santa Susana, California, 24 July 1959

Blockage of the coolant system in an experimental power reactor in the Greater Los Angeles area caused 12 of the 43 fuel elements to melt. The radioactivity was contained.

Cause: Blockage of coolant system

USA Enrico Fermi Reactor, SE Michigan, 5 October 1966

A malfunction in an demonstration fast breeder reactor 30 miles from Detroit caused part of the core to melt. Radioactive gases were contained in the reactor building. The plant was closed in 1972.

Cause: Malfunction

Switzerland Lucens Reactor, 21 January 1969

Pressure tube burst severely damaging core, rupturing the calandria and releasing radioactive coolant and heavy water into the containment. The reactor was built in a cavern under a hill and when the tube burst, automatic interlocks sealed the reactor cavern preventing the escape of radioactive materials to the surroundings. Reactor written off and cavern used for storage of radioactive materials.

Cause: Rupture of pressure tube.

USA Dresden II, Morris, Illinois, 5 June 1970

The reactor went out of control for two hours after a meter gave a false reading. Radioactive iodine accumulated in the containment vessel at 100 items the permissible level, but was not released to the environment.

Cause: False meter signals

USA Monticello, Minnesota, 19 November 1971

When the reactor's waste storage space filled to capacity, more than 50,000 gallons of radioactive waste water spilled into the Mississippi River. Some was taken into the St. Paul's water supply.

Cause: Waste-storage overflow

USA Brown's Ferry, Decatur, Alabama, 22 March 1975

A worker using a candle to check for air leaks around electrical cable set fire to insulation and destroyed cables controlling key safety equipment. Operators resorted to a makeshift system to cool the reactor core. Cleanup costs were $150 million. There was no release of radioactive material.

Cause: Human error

UK Hinckley Point B, June 1977

Major cooling pipe fractured and station staff had to rig a fire hose to keep the temperature of the concrete pressure vessel at safe levels. No release of radioactive material.

Cause: Pipe failure

UK Hunterston B, September 1977

Station staff rigged a makeshift pipe in the cooling system and forgot to remove it. When the reactor was depressurised for maintenance the makeshift pipe delivered several thousand gallons of raw sea water into the stainless steel core of the reactor. The reactor was shut down for many months and the total cost of the error was $70 million. No release of radioactive material.

Cause: Human error

USA Rancho Seco, Sacramento, California, 28 March 1978

A technician replacing a bulb behind a lighted push-button on the control panel dropped the bulb into the control panel where it shorted one of the main electrical connections in the reactor's instrumentation. Spurious signals about pressure, temperature, water level and flow flooded the main control computer which attempted to match equipment output to the erroneous signals. The computer shut off the feed water to the steam generators, the reactor pressure shot up and the reactor tripped. No release of radioactive material.

Cause: Human error

USA Sequoyah I, Tennessee, 11 February 1981

Eight workers were contaminated when 110,000 gallons of radioactive coolant leaked into the containment building. The accident threatened to expose the core of the reactor.

Cause: Coolant leak

Ginna Plant, Rochester NY, 25 January 1982

A major steam-generator tube ruptured spilling radioactive water into the containment vessel. Some radioactive steam escaped into the air.

Cause: Rupture of steam-generator tube

USA Salem NJ, 22 and 25 February 1983

Automatic shut-down system failed twice after water levels in the core dropped. Plant operators were able to shut down the core manually.

Cause: Failure of automatic shut-down system

USA San Onofre Unit 1, California, 21 November 1985

Unit 1 lost AC electric power for four minutes, leading to loss of coolant water. When water flow began again water hammer damaged plant equipment. Two workers received minor radiation doses. Plant still closed as of May 1986.

Cause: Equipment failure

USA Sequoyah I, Tennessee, 19 April 1984

A gusher of superheated radioactive water erupted during a maintenance operation. No radioactivity was released and eight workers in the area escaped injury.

Cause: Leak of pressurised water

USA Rancho Seco, California, 26 December 1985

In an echo of a previous incident the reactor suffered a loss of direct current power in its control system setting off a safety system which led to rapid overcooling. The incident did not damage the reactor vessel but "under alternate scenarios more serious consequences could occur" (NRC).

Cause: Weaknesses in the control system, management failure

USA Indian Point 2, Buchanan, NY, 28 December 1984

The reactor started up with all three safety injection pumps disabled. The pumps, a major part of the emergency cooling system, were blocked by solidified boric acid, a chemical that would, in an emergency, help shut down the reactor. A spokesman for the company said that at the stage the failure was discovered it was not a hazard.

Cause: Human error

USA Davis-Besse Plant, Oak Harbor, Ohio, 9 June 1985

At least 16 pieces of equipment failed and plant operators pushed the wrong button at least once. Although the incidents were similar to those that confronted the Three Mile Island operators, auxiliary coolant pumps were started and damage to the fuel core was apparently averted.

Cause: Equipment failure and human error

Public Service Electric and Gas Co. New Jersey, Hope Creek Plant, August 1986

Permission for loading of fuel and low-power tests given on 11 April 1986. Permission for full-power operation granted on 21 July and licence issued on 25 July. On 8 August it was discovered that the pressure equaliser had been installed backwards so that safety valves which should have operated automatically had to be operated manually. It was then found that the circuits had been installed according to the drawings and the drawings were wrong.

Cause: Human error

Most of these incidents have taken place in light-water reactors in the USA but that should not be taken to mean that nuclear operations there are less safe than in other parts of the world. There are more reactors operating in the United States than in any other country and public policy demands disclosure of any incidents, so that it is rather easier to obtain data on US accidents than in some other countries. The wide variety of causes of incidents fall either into human error or equipment failure and show vividly how the most trivial and even bizarre human failing can have wide-reaching, expensive and sometimes potentially dangerous consequences. This does not mean, however, as some environmental action groups have argued, that nuclear energy is too dangerous to meddle with but rather that it can be used with some confidence provided that the standards of construction are high and that the operators are trained to a high degree of competence and maintain continued vigilance such as is demanded in other industries such as transportation, petrochemicals and pharmaceuticals.

Chapter Nine

World Energy Use and Needs

The outstanding factor in the development of the human race over the last few hundred years has been its increasing use of energy. Increasing standards of living, improved health care and prolonged expectation of life are all primarily due to the increased availability of readily usable and cheap energy. It is a trendy feature of the modern environmental cult to look back longingly at the good old days in which everyone led a simple and natural life. The reality of the simple and good life can be seen in a visit to any country churchyard where the dates on the gravestones will show how short that good and simple life was for most people and how especially hard it was on women.

There were, of course, people who lived to a ripe old age, and very often survived three or four wives. They were the tough ones, and it is nature's way that the fittest should survive. In modern society we are no longer prepared to accept that the weak should perish and many of us alive today would not have survived, for not all of us are "the tough ones." Our lives have been prolonged by the existence of anaesthetics, antiseptics, vaccines and antibiotics, by adequate heating and clothing in winter and by the ready availability of transport to carry surplus food to areas of shortage — all products of a high-energy-use society.

The rate of consumption of energy varies across the world. Some of the reasons are obvious — some of us live in cold climates — but there is an almost universal truth, which is demonstrated in Figure 21, the higher the per capita use of energy in a country, the higher its gross national product and hence its standard of living. The relationship between these is almost linear. At the bottom of the scale is India with a per capita usage of energy of about 200 watts and at the top are

Canada and the USA with a per capita usage of more than 10 kilowatts and the highest standard of living in the world. These figures are not just electricity, but the total usage of energy in a country divided by the population.

There is one anomaly in the picture — that of Sweden which, with a climate not too different from that of Canada manages to achieve nearly as high a standard of living but with the expen-

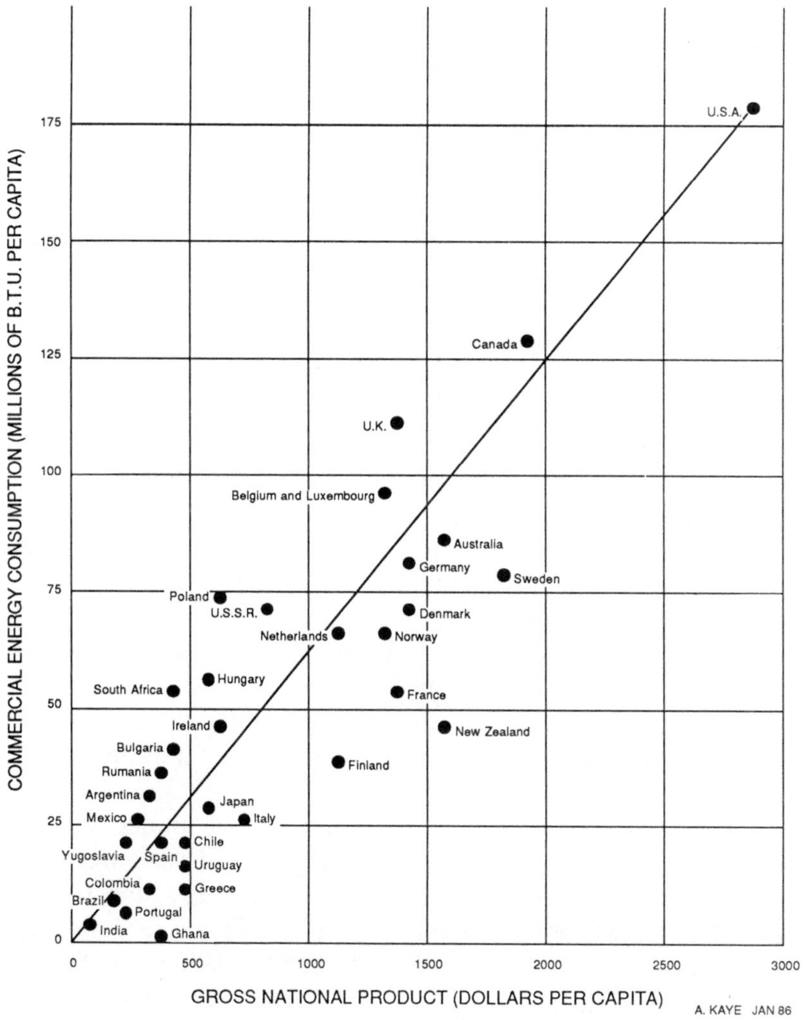

Figure 21 Graph of Energy Consumption vs Gross National Product

diture of about 65 per cent of the energy per capita. There are reasons for this, of course. In Sweden the population is more concentrated and distances for transportation are therefore not so great, so less energy per head is used in transport. However an important reason seems to be that real energy costs in Sweden have been traditionally higher than in North America and that has encouraged energy conservation to a much greater extent.

There is therefore some scope for further conservation in North America — smaller and more fuel-efficient cars, improved insulation and sealing of houses, although that too has its dangers — but the fact remains that in North America energy use will be unlikely to fall much below the 10-kw-per-capita level. When you consider that a healthy man, working hard, does work at about the rate of 500 watts you realise that everyone in North America is using energy as if each had twenty healthy slaves. And this is perhaps the greatest achievement of the human race in the last two hundred years — to replace the switch of the overseer by the electrical switch and control valve.

The increase in the use of energy has been quite phenomenal since 1900. It is hard to believe that 99 per cent of the fossil fuel which has been used since the beginning of time has been burned since the year 1900 and 50 per cent since 1945. Thus our use of energy since 1900 has not only been increasing, but increasing at an increasing rate. For the thirty years after World War II annual electrical demand in the western industrialised world increased at a rate of seven per cent. This implies a doubling time of ten years and demands that in any ten-year period the new generating capacity to be built should equal the total generating capacity in existence at the start of the period. It was obvious that such growth could not be sustained forever, but not easy to forecast when it would fall off.

Electrical generating authorities have come under a lot of criticism lately for building excess capacity and not forecasting that the growth rate would fall to its present less-than-five per cent. The critics seem to forget that the gestation time for a large generating station is between 11 and 12 years. Whatever the level of criticism of the overestimation of growth in demand, it is as nothing compared to the outcry of rage that would have erupted had the error been in the other direction! It seems likely that there will not be a great reduction of per capita energy usage worldwide in the future, rather, indeed the opposite when the developing nations realise (as some have already) that their economic wellbeing is tied to a greater and greater use of inanimate energy. Taken in conjunction with the fact that the world's population is still increasing (the five-billionth person was born in July 1986), it seems obvious that, rather than world energy consumption being reduced in the future, it will go on increasing. We will thus be faced in the future not only with an increasing demand, per capita, for energy but with an increasing number of people making the demand.

To have some idea of how the world's resources can meet the demand we have to make some estimates of population growth. Let's err, if at all, on the optimistic side and assume that the world's population will stabilise at eight billion (8×10^9) and that each of these people uses energy at the rate of 10 kilowatts. This is about the same as the current North American rate. A rate of ten kilowatts means 10 kilowatt hours per hour or 240 kwh per day, corresponding to an annual rate of 87600 kilowatt hours per person. The 8 billion people in our steady state world would thus use per year

$$8 \times 10^9 \times 87600 = 7 \times 10^{14} \text{ kilowatt hours}$$

This is already an inconveniently large number so we will use another unit, the Q, which is equal to 300 million million (3×10^{14}) kilowatt hours. We thus arrive at the annual energy consumption of our stabilised world (each person in which uses energy at the rate of 10kw) of 2.3Q.

Can the resources of the earth sustain this demand? We can only answer this by trying to estimate the energy resources of the earth. Some of the resources can be determined quite accurately but others are much more uncertain. No one really knows how much coal, oil and natural gas there are buried in the earth; we can only look at the proven reserves and study how the rate of discovery of new reserves has changed over the past decades. From that we can arrive at an estimate of the total resource, which may be inaccurate by as much as a factor of two.

However that does not really make much difference to the final analysis in the long term. In a hundred years' time it will not be of much significance whether the oil lasted until 2020 or 2050 AD!

World Energy Resources

Fossil Fuels

World energy resources from fossil fuels are summarised in Table 8, due to Thompson (1977). There is about 190Q of fairly readily accessible coal. If our stable world used nothing else but coal it would be sufficient to last for about eighty years, and if we were prepared to mine more difficult to get at coal (with less than 4000 feet of overburden and in seams greater than one foot thick), there is about twice as much. Surely enough time, one would think, for some more satisfactory system to be developed. There are compelling reasons, however, why the use of coal should be decreased rather than increased. It is dangerous to mine — the cost of a hundred years of coal mining in North America from accidents alone is 100,000 lives and in Britain 50 miners are still killed every year in accidents. In addition many miners are disabled and die prematurely from pneumoconiosis and black lung.

However it is when we start to burn coal that it really gets dangerous. Even under optimum burning conditions each atom of carbon burned results in the formation of a molecule of carbon dioxide which is released to the atmosphere. As a result of the burning of fossil fuel the carbon dioxide content of the atmosphere has been increasing steadily since the year 1900. Carbon dioxide has the peculiar property of allowing short-wave radiation from the sun to reach the earth unscathed, but of trapping the long-wave radiation emitted from the earth. Increasing carbon dioxide content of the atmosphere will thus result in a warmer atmosphere and there is already evidence that this is occurring. Many scientists are therefore concerned that further widescale use of fossil fuel may result in a temperature change

Table 8
World Energy Resources

Fossil Fuels	**Units of "Q"**
Coal	190
Coal (in seams > one foot thick, < 4000 feet deep)	400
Petroleum	11
Natural gas	10
Tar sands	1.7
Oil shale (proven)	1.1
Oil shale estimated at 25 gallons/ton	100
Nuclear	
Uranium (proved)	1.0
(if used in breeding cycle)	80
Uranium (estimated at < $130/kilo)	2.2
(if used in breeding cycle)	160
Thermonuclear	
Deuterium-deuterium reaction	effectively limitless
Renewable (quantity of energy available annually)	
Solar (at surface of earth)	800
Wind (usable)	0.1
Hydro	0.2
Tides (total)	0.1
Geothermal heat flux	0.8

(Units of "Q," 1 Q = 3×10^{14} kilowatt hours)

(Source Thompson, 1977)

sufficient to melt a large proportion of the polar ice caps, raise sea levels by considerable amounts and cause a potentially disastrous climate change over much of the earth's surface.

This is by no means the end of the story. The other substances emitted when coal is burned are summarised in Table 9. Sulphur dioxide emitted from coal-fired generating stations is one of the main causes of acid rain which has killed thousands of lakes in Scandinavia, Germany and North America. The other pollutants emitted, the aldehydes, are even more noxious and finally, and perhaps surprisingly, all coal-fired stations emit radium, admittedly in small amounts.

Table 9

Typical Pollutant Emissions from Non Nuclear Generating Stations of 1000mW(E) Capacity

	Tonnes per year		
	Coal	**Oil**	**Natural Gas**
Aldehydes	5	117	31
Carbon monoxide	520	8	0
Hydrocarbons	210	670	0
Nitrogen Oxides	21000	21700	12000
Sulphur Oxides	139000	52600	12
Particulates	4500	730	460

In addition, the coal-fired station will emit about 30 millicuries of radium per year.

If coal were first discovered tomorrow it is probably no exaggeration to say that it would be banned without further ado as a dangerous substance, highly detrimental to health and the environment. Apart from anything else, a large coal-fired generating station produces in a single year enough ash to cover an acre of ground to the height of a six-storey building. The ash itself contains carcinogenic and mutagenic substances. It is expected that the combustion of coal in fluidised beds will give up to 20 per cent more energy due to increased efficiency of combustion and will also produce fewer pollutants. The general application of fluidised bed technology is probably still 10 to 20 years away. It is possible to produce synthetic oil from coal but the cost is very high, more than $35 per barrel and the total production of oil from coal in the future is not expected to be large.

The situation with oil and natural gas is very much more serious. There is only about 10 or 11Q of each — sufficient to keep our equilibrium world going for only four or five years, if we used nothing else. Our present use of energy is only about one tenth of this but even that is increasing exponentially, admittedly at a slower rate than a few years ago, and it might be expected that supplies of oil and gas, at our present rate of usage, might last another 40 or 50 years. But that is only if the developed world continues to take the lion's share of energy, and, even if it wanted to that will not continue to be possible. It is ironic to hear the plaintive cries of the newly-oil-rich nations at the falling price of oil in 1987 and to remember that the rise in the price of oil which made their fortunes ten years ago, proved only an inconvenience to the developed world but was a catastrophe to the developing countries.

Of course, it will be argued, more oil and natural gas will be discovered, and that is true. But it will be discovered at a decreasing rate, the rate of use of oil and gas has long since outdistanced the rate of discovery of new reserves. The new reserves when discovered will be much more difficult and expensive to tap — the easy oil and gas is already being exploited. In the 1980s we have already started to use the difficult and expensive reserves and this has only been made possible by the price rises imposed by OPEC in the 1970s. The 1986 collapse of oil prices has already practically caused a standstill in new exploration and shut down many oilfields which only oil at $35 a barrel made economic to operate.

Since oil is the almost universal fuel for mechanical transport and will become increasingly scarce, it is rather silly to burn it to generate electricity. Similarly natural gas is also a scarce resource and it seems a pity to use it to generate electricity at an efficiency of perhaps 40 per cent when it can be burned in home furnaces at 75 per cent efficiency. What other methods are there for generating electricity which, if industrial expansion is to continue, will be the form in which most "new" energy will be required?

Coal, oil and natural gas are the results of energy from the sun which was stored in plant life some 600,000 years ago. They are therefore not renewable on a human time scale. Forests and plants store solar energy on a much shorter time scale and are therefore "renewable" energy sources. Most poor countries depend on wood for energy and nearly half of India's energy and more than 90 per cent of Malawi's comes from wood. That is one of the reasons why those countries are poor. Log fires are pleasant as a luxury, but not if you have to depend on them for all your heating and cooking. There has been a recent vogue in wood burning stoves in North America but they are expensive to buy and to run (with wood cut and transported at North American wage rates) and have also caused a number of deaths due to the production of carbon monoxide from an insufficient supply of air for complete combustion.

In addition the 12 million wood stoves in use in the United States are believed by the Environmental Protection Agency to produce about 15 per cent of the national emission of small air-borne particles and 40 per cent of the emission of certain chemicals (including several carcinogens) created in fires which are starved of oxygen. From 1987 the EPA will propose emission limits for wood burning stoves on similar lines to automobiles and factories. As a result many new stoves will have to be equipped with catalytic converters. Local areas all over the United States have imposed regulations on such stoves and, for example, Telluride, Colorado will issue a permit for a new stove only if the applicant can persuade two other residents to give up theirs. In October 1986 the town began offering a bounty of $750 to anyone who replaced a wood burning system with natural gas. Already in the world 1.3 billion people who depend on wood for fuel are using up trees faster than they can be replaced and it is unlikely that wood will increase its share of energy production.

In India about ten per cent of all energy is derived from burning animal dung. This is a gross waste of a valuable resource and is the start of a vicious circle, for farmers, in burning dung, deprive their fields of fertiliser and thus are able to grow less crops which, in turn, provide less food for animals which would produce the dung. A more sensible use for dung would be to allow it to decompose anaerobically (i.e. out of contact with air) to produce the gas methane which is an excellent fuel either for heating or (like propane) for fuelling automobile, tractor or truck engines. The anaerobic decomposition of manure leaves a residue which is still a good fertiliser. China is already producing the annual energy equivalent of 20 million tonnes of coal from simple backyard manure digesters. There is also a gentleman in England who enjoyed some media publicity a few years ago when it was revealed that for twenty years he had been running his car on methane produced from chicken manure. However while such a system may well improve the lot of the Indian peasant it is unlikely to contribute significantly to the required 2.3Q per year.

Renewable Energy Resources

What then about the truly renewable energy resources? The first among these is the sun — an enormous thermonuclear reactor turning mass into energy at the phenomenal rate of four million tonnes per second, what's more, at the fairly safe distance of 93 million miles from us! It has been producing this energy for billions of years and will go on doing it for billions more. Every facet of our lives depends directly or indirectly on solar radiation. About 5000 Q per year arrives from the sun at the outer edge of the earth's atmosphere and about 800 Q penetrates to the surface, plenty to more than satisfy the now seemingly modest demands of 2.3 Q which our equilibrium world will make. Attempts to harness this energy have been made, with varying degrees of success, since the beginning of time. Solar water-heating was almost universal in Florida in the 1930s and only fell into disuse with the arrival of convenient and cheap electrical energy. The increase in energy prices in the 1970s spurred a great deal of interest in solar heated homes, in which solar panels are mounted on the roof, the sun's energy is used to heat water which is circulated and stored. Various demonstration schemes were carried out and worked rather well. The problems are that solar radiation is intermittent and that any building which uses it for heat must have a storage system (either a large tank of water or a bed of rocks) in the house to store perhaps two or three day's energy needs. This adds considerably to the cost of a solar heated house. The technology is also rather expensive at present, precisely because there is no mass market and there will be no mass market until the cost falls. It is a fact of life that the last ten thousand dollars to put towards the purchase of a dream house is always the hardest to raise and the prospect of saving perhaps a thousand dollars a year in heating costs is not usually a sufficient incentive for a house purchaser, already probably rather strapped for money, to lay out an extra ten thousand dollars when buying his house.

Central Solar Systems

Using solar energy for direct home heating is just the most obvious way of utilising it. Other schemes have been proposed for solar operated electricity generating stations and some demonstration installations are already working. In Israel, for example, there is an experimental 5-megawatt generating station based on a brine-filled pond and in California a 14-megawatt installation based on parabolic mirrors in the form of troughs. They work very well but they are expensive, up to $2000 (at 1986 prices) per installed kilowatt. It is unlikely that they will contribute significantly to our equilibrium world's energy needs, except on rather a small scale.

Photovoltaic Cells

Sunlight can be turned into electricity directly using semiconductors. However they are rather inefficient converters and, while they are very useful in electrical "gimmicks" such as solar powered calculators requiring a few milliwatts of power, or in serious applications such as satellites where cost is not the controlling factor, they are rather impractical and expensive for large scale energy generation. Even with another ten years' development it is expected that the cost of electricity generated from photovoltaic cells will be around 23 cents per kwh (in 1986 cents). Their big advantage lies in isolated areas which are a long way from a generating station, for they can be made as small or as big as desired. At present about 20 megawatts of photovoltaic capacity is connected to the grid in the USA. In ten years time the figure could have grown to several thousand megawatts.

Another problem is that, while the total amount of energy arriving at the earth's surface is huge, it is spread over a lot of surface so that the power density is low. The amount of land required for a solar generating station is thus rather large. A 1000 megawatt solar station might occupy 5000 acres as compared to the 150 or so of a conventional or nuclear station.

Hydroelectric Power

Using hydroelectric power is one way of tapping solar energy; the sun evaporates water which eventually turns into clouds, some of these result in rain and some proportion of that rain is deposited at a high location so that it can be allowed to run through turbines and generate electricity. A good deal of Canada's electricity is generated by this means, for example Niagara Falls and the James Bay Complex. The trouble is that the total amount of hydroelectricity is not large in relation to the needs we are discussing, about 0.2Q, and most of the readily accessible and usable hydroelectricity has already been tapped. There is, however, a great deal of scope for the development of fairly small scale hydroelectric sites and many electricity authorities, Ontario Hydro for one are actively examining the possibilities.

Wind-powered Energy

No one who has struggled to bicycle against the wind will doubt that it contains energy and this is one of the oldest sources tapped by man, originally for grinding corn and then for pumping water and latterly for generating electricity. The wind is free, its energy also comes from the sun and it is largely a vast and unexploited source. It is estimated that the energy available from the wind amounts to about 100Q per year, very much more than is required to fuel our stablised world. Unfortunately there is a snag. Most of the wind's energy is available only in areas where it is not needed and only about 0.1Q per year is harnessable where it will do most good. Single small windmills can be very successful but attempts to generate large quantities of energy by scaling up windmills have proved disappointing.

At the present time much of the world's installed capacity is in California where the tax system favours windmill owners. At Altamont Pass in California about 7,000 windmills have a peak capacity of 1200 megawatts (about the same as a large generating station), but they occupy nearly 17,000 acres of land. The capital cost is high — about $1500 per installed kilowatt and the cost of electricity generated by wind energy is not expected to fall much below ten cents per kilowatt hour. Wind energy will continue to be useful on a small scale and in specific areas. It is unlikely however to make more than a few per cent contribution to the energy needs of our stabilised world.

Geothermal Energy

The core of the earth is molten and in certain places the crust is so thin that some of the heat can be tapped at the surface of the earth. This is not strictly a renewable source of energy since it is not being replenished, but nevertheless there are large amounts of energy waiting to be used. They are tapped to some extent already in such well known cases as Iceland and New Zealand. Not so well known perhaps is that there is already a geothermal electrical generating station operating in California at about 300 Megawatts. We can estimate the rate at which this energy could be tapped by measuring the rate at which heat escapes through the crust of the earth. As a result of many of these measurements it is known that the heat flux through the surface of the earth is about 0.1Q per year. Even if a large part of this could be harnessed it would not make much of a contribution to the energy needs of our equilibrium world. It would be possible to bore down and extract heat from hot rocks, but, even then the total amount of energy in the outer few kilometres of the earth's crust is only about 100Q and tapping even a few per cent of that might cause major instabilities.

Tidal and Wave Energy

No one since the days of King Canute has doubted the power of the tides and the idea of harnessing this energy has preoccupied engineers for a century. In principle it is very simple to do so, not any different from impounding a lake behind a dam and generating hydroelectricity, with the added advantage that the lake fills up twice a day automatically and one doesn't have to wait for rain. There are two constraints, one needs a place where there is a large difference in height between high-water and low-water levels and a long river estuary. Once these have been located all that is necessary is to build a dam across the estuary and let the tide pass through the dam when it is rising. When the tide goes out the impounded water is allowed to fall through turbines. The snag is that there are probably only twenty or thirty suitable sites in the world, including the Severn Estuary in Britain and the Bay of Fundy between Nova Scotia and New Brunswick. There is already a successful 300MW(e) tidal power station operating on the estuary of the River Rance in France but the total tidal energy itself is relatively small, only about 0.01Q.

The ocean waves contain large quantities of energy, about 90 kilowatts per metre of wave front in mid-Atlantic and many ingenious devices have been proposed to harness this energy. The best known are the Lancaster Bag, the Salter Duck the SEA Clam and several based on the oscillating water column. There was a flurry of interest in wave power in the early 1980s but this seems to have died down mainly due to the projected cost of around 20 to 25 cents per kilowatt hour and to the probable difficulty of ensuring structural integrity. Norway has launched two small wave-power stations on an island near Bergen and, while at present it has sufficient electricity from hydroelectric sources, it is prepared to market them.

A device invented by Vickers in England consists of an 80-metre tube weighing 20,000 tonnes and driven into the seabed. The rise and fall of the waves causes water to rise and fall within the tube and generates air pressure which drives a turbine and generates electricity. A 2000 Mw(e) generating station would require 700 such tubes stretching over 60 kilometres of seabed. The big advantage over other devices is that the Vickers tubes would be in deep water below the surface and would thus be free from the structural stresses suffered by floating units. It is claimed that electricity could be generated for about 10 cents a kilowatthour by these devices.

Nuclear and Thermonuclear Energy

The world's proved uranium reserves which can be recovered at a cost of less than $130 per kilogram amount to about six million tonnes. If $130 a kilogram sounds rather a lot to pay for fuel, remember that a kilogram of uranium is equal in energy terms to about 200 tonnes of coal, and 65 cents a tonne is rather a good price for coal. In energy terms, six million tonnes of uranium, if used in PWRs at 70 per cent load factor would generate a total of about 1Q, enough to fuel our equilibrium world for about six months. If we use the figure for estimated resources, as opposed to proven, the quantity of energy goes up substantially, but is still only comparable to oil. On the other hand if the uranium was used in breeding cycles it would generate about 80Q. From any point of view therefore, nuclear energy can never be regarded as more than a stopgap source. Something that may be useful for perhaps fifty or a hundred years so that the time gained can be utilised to develop a renewable source for our equilibrium world.

We release nuclear energy by causing fission in atoms of uranium. At the other end of the atomic scale we can release thermonuclear energy by causing atoms of light elements such as hydrogen, deuterium and tritium to come together in fusion. There is nothing new in this, the sun has been doing it for a long time. The difficulty is that the atomic nuclei are positively charged and have a mutual repulsion. To make them fuse we have to make them go very fast indeed in order to overcome the repulsive force and that entails raising them to very high temperatures of around a hundred million degrees. Unfortunately there are no materials on earth capable of containing a gas at this temperature, so the hydrogen molecules have to be kept away from the walls of the container by means of a magnetic field. The success of controlled thermonuclear energy will ultimately depend on the success of the physicists in designing magnetic fields which are capable of confining the hot gas long enough for the reaction to proceed. If that is achieved the rewards may be almost limitless for the sea contains a great deal of deuterium — one molecule of sea water in 6,000 contains a deuterium atom. The problem is the time scale on which success will be achieved. If one had asked a physicist in 1950 when we would see the commercial application of thermonuclear energy he would have replied "in about 30 years" and would have considered that he was being cautious. If the same question were asked today the answer would be the same.

The Future

Unless there is a considerable change in the present pattern of energy consumption it looks as if the major part of the world's future demand for energy will be in the form of electricity. All of the renewable forms of energy discussed above with the exception of biomass, can produce electricity conveniently if expensively. The total contribution of nuclear-generated electricity to the world's energy use is still quite small. According to the IAEA at the end of 1985 the total contribution of nuclear-generated electricity to the world's electrical supply was 15 per cent. However only about 35 per cent of the energy used at present is in the form of electricity and nuclear power contributes only about 5 per cent of the energy delivered to the world's users. Certain countries make much more use of nuclear energy than others — about 65 per cent of the electricity used in France is nuclear-produced. The growth of nuclear energy has been much slower in the last decade than was anticipated, mainly due to the drop in demand for electricity in the recession of the early eighties, but also, and in North America significantly, due to the aftermath of the Three Mile Island accident and the consequent increased attention to nuclear safety, together with the escalation in cost of construction. It is anticipated therefore that the installed nuclear capacity by the end of the century will be barely half that forecast in 1975, although double that installed capacity of 1986, i.e., about 500,000 Megawatts. If it were decided to phase out nuclear energy there would be a considerable shortfall, even without considering much increase in demand from the developing countries.

How would it be possible to fill the gap? The most obvious answer is by conservation, remembering that conservation does very little to fulfil the aspirations of the developing world. Conservation has certainly been effective in the last decade — fed by the increasing cost of energy — and statistics show that in the western developed countries the amount of energy used per unit of gross domestic product has been reduced by 13 per cent since 1974. The problem with continuing this process is that we come up against a law of diminishing returns which means that each further unit of energy saved will be at increasing cost. The easy and economical ways of conserving energy — improved home insulation, double glazing, improving the efficiency of furnaces have generally mostly been taken.

Suppose we decided to phase out nuclear power, what would replace it? Coal certainly could but, as has been shown, it is not an attractive alternative. If we wish to use more coal we either have to accept more pollution or pay more to clean up the coal before burning it. We also have to accept a toll in human suffering which falls on a selected group of people and not, as it should perhaps, on those who use the electricity generated from coal. If nuclear plants were to shut down overnight brownouts and blackouts would occur on a widespread scale and consumption of coal would have to increase by about ten per cent. It would be more practical, perhaps, to phase out nuclear power (if it were thought desirable), over the next twenty or thirty years, but a recent estimate by the *Economist* magazine suggested that, as a result, energy prices would at least double in real terms with a possibly disastrous effect, not on the western developed nations, but on those least able to afford it, the developing third world countries.

It has been argued that had equivalent money been directed to developing renewable energy sources as have been devoted to nuclear power, those sources would now be practicable and economic and nuclear power would be unnecessary. This argument overlooks the fact that the renewable resources have innate problems which are unlikely to be soluble in the short term — they require large areas of land and they are intermittent.

Chapter Ten

Risk and the Generation of Electricity

Most people would deny that they had any specialised knowledge of risk analysis, but nevertheless have an innate confidence in their ability to recognise and assess danger. Most of us would assert that we would only rarely, if ever, set out deliberately to do anything which we would consider dangerous and this implies that each of us has an inherent mechanism for assessing risk and that it is highly effective. The truth is rather different for some of the things which many people do every day without thinking are fairly dangerous and others which are widely regarded as being dangerous are really rather safe.

Our built-in risk assessment mechanism is intensely personal and in particular, our perception of whether a certain pursuit is dangerous or safe is strongly influenced by whether we derive some benefit from it, whether material or in the form of enjoyment, and especially whether we follow the pursuit or accept the situation by choice or whether it is forced upon us.

Risk analysis is generally regarded as a fairly new science but, in fact, in the form of insurance, it has a long and mostly respectable pedigree. Sir Nicholas Bacon, addressing the first Parliament of Queen Elizabeth I in the sixteenth century said, "Does not the wise merchant in every venture give part that the whole may be assured?" and, from the sixteenth century there is an extant contract between Daringo Simon Maiar and two women to whom he agreed to pay thirty livres on the birth of daughter on the undertaking that he receive forty-eight livres in gratitude for the birth of a son. In 1693 the Philosophical Transactions of the Royal Society of London published a table based on the births and deaths in the city of Breslau. The purpose of Halley's life table was an attempt to "ascertain the price of annuities on lives."

Since then, at least for most of the time, insurance companies have made handsome profits by forecasting the connection between risk and life. Risk analysis has now become rather a sophisticated science based partly on the frequency of past events from which it is hoped to derive reliable forecasts of the likelihood of such events being repeated in the future. The results are expressed in terms of probability, or the odds of a certain event happening. As such it inevitably carries some connotation of gambling which implies a degree of frivolity which may seem unseemly in connection with matters of life and death.

However, most people would agree that it is foolish to undertake any activity without being aware of the probability of success or failure — in other words, the odds. One big difference between risk analysis and gambling is the length of these odds. Most of us would regard a horse at a thousand to one as a longshot indeed, would think twice before backing it and would consider ourselves fortunate if it won. In risk analysis, odds of a thousand to one of death in a particular year due to a particular activity are very short odds. Even then, a thousand people get away with doing whatever it is, but for the one person who gets caught, it is no longer a risk, but a certainty and final one at that.

Our perceptions of risk change markedly with age. Most of the really risky things in life are done by young people and those who continue motorcycle racing, rock climbing and parachute jumping to advanced years are regarded as eccentric, to say the least. Most of the statistics expressing the risk of everyday activities are surprising. This can be illustrated by comparing the risks encountered in their normal living by two people with very different lifestyles.

One of them is a farm worker who leads a healthy open-air life, rides five miles to work every day on a motorcycle, smokes twenty cigarettes a day, plays baseball in the summer and ice hockey in the winter.

The other is a high-level businessman who leads a stressful life of meetings and travel by air. His business lunches start off with a martini and go on from there. Dinner is the same only more so. He admits to drinking the equivalent of a bottle of wine a day. He has a passion for downhill skiing. It's pretty obvious which of these leads the more dangerous life, or is it?

Let's deal with the lives of each, point by point. The healthy open-air life of the farm worker is, in reality, rather a dangerous one. Working on a farm carries a 1 in 9,000 chance of being killed in a year. He travels five miles a day on a motorcyle and has a 1 in 500 change of being killed in a year as a result. His 20 cigarettes a day give him a 1 in 200 chance of dying per year (averaged over the number of years he smokes). His recreations are rather safe, perhaps a 1 in 50,000 chance per year of being killed.

In contrast our business executive has a very safe job. His chances of being killed at work are probably less than if he stayed at home. His air travel of perhaps 30,000 miles per year carries a 1 in 10,000 chance of being killed in an accident, very much safer than going five miles to work on a motorcyle every day. His heavy drinking is not all that dangerous, perhaps a one-in-thirteen-thousand chance of dying in a year due to his drinking — not nearly as high as the danger to even a moderate smoker. Downhill skiing is, perhaps surprisingly, a comparatively safe sport and carries a risk of being killed of about one in 1,150,000 per hour of skiing. Our executive can therefore ski for about 23 hours per year without exceeding the risk of the farm worker's ice hockey and baseball.

Overall, therefore, the farm worker is dicing with death to a much greater extent than the jet-setting business executive and therein lies the first lesson of risk analysis, it is not always the humdrum, everyday activities which are the safest or the exotic which are the most dangerous. For example the risk run by a motorcyclist, riding only ten miles a day, of a one-in-five-hundred chance of death in a year is the same as the risk to a person living in a town of 10,000 people which has a nuclear reactor which blows up once every five years killing a hundred people. The point is that no one in their senses would go to live in a town like that, but many people, who are not otherwise considered insane, ride motorcyles.

The risks of travelling a million miles by various forms of transport are summarised in Table 10. It is not surprising that the motorcyle tops the list with a one in three chance of being killed followed closely by horseback riding and the pedal cycle. It is apparent that it is much safer to travel a given distance by air than it is by car which will come as a surprise to most people since fear of flying is quite common and fear of car driving is rare. The reason, of course, is that when an accident happens to an airliner many people are usually killed at once while traffic deaths (which in total are very many more than air crash deaths) are spread out over the year and over the whole country and may make headlines in local newspapers but cause scarcely a stir nationwide, in spite of the fact that road deaths in the United States have, for many years, averaged close to 50,000 per year.

Table 10

Risk of Death in Travelling a Million Miles

Motorcyle	1 in 3
Horse	1 in 4
Bicycle	1 in 6
Car	1 in 66
Airplane	1 in 300
Train	1 in 300
Bus	1 in 500

One can, of course, protect oneself to some extent against most risks by taking safety precautions. For example the risks of motorcycle riding can be reduced by wearing crash helmets and those of car driving by wearing seatbelts, keeping to speed limits, taking note of road hazard signs and even by the ultimate precaution of travelling everywhere by bus. It is important to realise that, while safety precautions can and do reduce risks substantially, they never eliminate them completely. The most prudent car driver, sober, seatbelted and within the speed limit is still at the mercy of a drunken driver ignoring traffic lights and driving at a hundred miles an hour. Even the committed pedestrian puts himself at risk every time he crosses the road.

Risk in Energy Generation

Most of us pay our electricity bills and consider that that is the end of the matter. So it is for us, but just as there is no human activity without risk, so there is a social cost to electrical energy in addition to a monetary one. We each pay the cash cost in proportion to the amount of electrical energy we have used. Who pays the social cost? So far we have been considering risk on a more or less personal basis. Can the same sorts of arguments be applied to the risks of living in a high-tech environment, to the coal mines, the chemical factories and even the nuclear reactors? They can and are, sometimes, with the same surprising results. Most people would say that the safest way of generating electricity is by a hydro-electric dam. After all, no coal mining is involved, no burning of fuel and no pollution, and certainly no radioactivity. On the contrary, dams are relatively dangerous things. When we look at the statistics of hydro-electricity we find that in July 1985 a burst dam swept down the Fiume valley in Northern Italy killing 260 people; in 1979 in India, fifteen-thousand people were killed in the collapse of a dam and in 1963 nineteen-hundred were drowned at Vaiont, Italy in a similar accident. Strangely enough none of these terrible events was followed by mass demonstrations advocating the banning of hydroelectricity or a public outcry calling for the emptying of dammed lakes and the demolition of the dams.

The generation of any energy will always carry a cost in human suffering. Lumbering is a dangerous occupation, the transport of logs results in casualties, the burning of wood causes many deaths every year, especially in cold climates, both by carbon monoxide poisoning and by fire. The generation of electrical energy will also cause casualties by whichever method it is done. The only feasible way to phase out nuclear generation within the next twenty years or so, if society should decide that it is desirable to do so, is by an increasing dependence on the use of coal to generate electricity. It seems sensible therefore to try to consider and compare the social cost of each method.

Coal-fired Electricity Generation

At first glance the accident statistics for coal mining are appalling. The annual death toll in accidents in US coal mines between 1960 and the present varied between 100 and 300 and in Britain about 50 miners are killed in accidents in coal mines every year. The output of a single large modern generating station is about 1000 Megawatts (electrical) and a coal-fired plant of this size would burn about 3.8 million tons of coal per year. The most careful assessment of the risks of mining coal for power stations has been carried out by Ferguson (1981) for British generating stations. He identified the coal mines which supplied the generating stations and used the accident statistics for those mines to determine the number of casualties to be expected among the miners. He arrived at a figure of 0.7 deaths per year per 1000 Mw(e) generating station. To accommodate the variation in the accident rate from year to year, Ferguson derives limits of from 0.3 to 1.7 deaths (in coal mining) per 1000 Mw(e) generating station per year. The best fit to many years of accident statistics in the UK gives 1.3 deaths per year for the mining of 3.8 million tons.

The risks involved in burning the coal to generate electricity are not so easy to quantify and the estimates of various authors vary rather widely. The total fatality rate to both miners and the general public from mining and burning the coal ranges from 1.9 to 11.8 deaths per year (for a single 1000 Mw(e) generating station) depending on whether one takes an optimistic or a pessimistic view of the risk of fatal disease to members of the general public as a result of the gaseous emissions from burning coal. These risks, whether they are at the lower or upper end of the estimates are shared by a selected group of people, the coal miners and those people who live downwind of large generating stations. The majority of us are fortunate enough to be able to pay for our electricity with money alone and are not asked to contribute significantly to the social cost.

The other deleterious effects of electrical generation in coal-burning plants are environmental — the addition of carbon dioxide to the atmosphere contributing to atmospheric warming by the greenhouse effect and the major contribution made by coal burning to acid precipitation. These effects are certainly real, they know no frontiers and are difficult to quantify. They can only be eliminated by national willingness to pay the price and by international agreement.

Risk in Nuclear Generation

The risks involved in the early years of uranium mining were also very high and it is reported that 50 per cent of the uranium miners who worked in the Joachimstal mines in Czechoslovakia eventually died of lung diseases from the inhalation of radon daughter products in the mines. Even in the immediate post World War II years in North America, conditions were far from ideal and it was only in the 1960s that radon levels in the uranium mines were brought down to acceptable levels. Since there is a latent period of 10 to 20 years for radon-induced cancers, it is only in recent years that the death rate from this cause has fallen to very low levels. Other diseases which have traditionally affected uranium miners, such as silicosis have also been virtually eliminated by improved working practices. The last uranium miner to claim compensation for silicosis in Ontario commenced his compensation in 1981, and his disease had probably been contracted over the previous twenty years.

In practice uranium miners have a lower than average standard mortality rate for death from disease, but an extremely high mortality rate from violent causes such as car accidents (Muller et al, 1984), which apparently is related to life style in "boom-bust" uranium towns. The total quantity of uranium required to fuel a 1000 Mw(e) reactor for one year depends on the reactor type, fuel burnup and fuel recycling but is between 50 and 300 tonnes. Depending on the grade of the ore between 50,000 and 300,000 tonnes of ore will have to be mined. At the maximum this is less than one-tenth of the quantity of coal required and, if it is assumed that fatal accidents are a linear function of the total tonnage mined, the cost in the mining step of running a 1000 Mw(e) reactor for a year will be at least ten times greater for coal fired electrical generation than for nuclear. More sophisticated calculations put the present day figure at 26 and predict that it may rise to 70 in the next 50 years.

There are additional risks to the general public from the detritus of mining operations. A sliding slag heap (from a coal mine) swept a school away and killed more than a hundred children at Aberfan in Wales in the 1960s. The risk to the general public from uranium mining arises from the radon released from old mine workings, from the milling operation and from covered mine tailings. These have caused considerable concern to environmental groups for the release of radon from the workings and from covered mine tailings will effectively continue forever. It has been shown by Branagan and Gotchy (1981) that the uranium required to run a 1000 Mw(e) station for one year would result in a percentage increase of the North American background radon concentration of less than four millionths of one per cent for the mining operation and one thirtieth of that level for the milling and covered tailings. These are very small increases and when multiplied by the population and by currently accepted risk factors might result, over a period of 500 years, in 0.2 additional deaths.

The results of 10 major surveys of comparative risk of operating a 1000 Mw(e) station for one year are summarised by Dunster (1981) as:

coal	0.6 - 10 deaths per year
oil	0.1 - 1.5 deaths per year
nuclear	0.1 - 1 deaths per year.

If recent American views on the effects of exposure to low concentrations of sulphur dioxide are accepted, the upper limits for both coal and oil would rise to about a hundred deaths per year. As Dunster says, a generating station of this size would serve a population of about a million people among whom 14,000 deaths would normally occur per year. Although the deaths associated with the generating station do not make a large difference to the overall mortality of the area in which the station is placed, avoidable extra deaths are not acceptable. The figures do suggest, however, that if society's overall objective is to reduce the number of premature deaths, there may be more effective ways of doing it than by concentrating on the deaths connected with the generation of electricity.

Accidents to Nuclear Reactors

As a result of the foregoing analysis it has been established fairly definitely that, as far as mining the fuel and normal operation of the generating stations is concerned, the generation of electrical energy is very much safer and more environmentally acceptable by nuclear methods than by coal fired. It now remains to consider accident conditions. When you read the descriptions of accidents given in Chapter Eight you might be forgiven for thinking that nuclear reactors are devices designed by clowns and operated by morons. Nothing could be further from the truth. Reactors are designed with safety first and foremost in mind and contain many engineered features designed to deal with any conceivable fault. The basic philosophy of the US Nuclear Regulatory Commission is typical and will be considered in some detail. The basis of the philosophy is that there are three levels of safety designed into the reactor.

The first level is to design the reactor and its associated equipment so that they will work very reliably and have only a small chance of a failure occurring. Special emphasis is placed on the quality of materials and workmanship. The plant must be designed to withstand hurricanes, tornadoes and other natural phenomena, and to be capable of a safe shutdown in the event of the worst earthquake which might be expected at the plant site. The design must permit continuous monitoring of components and systems to detect signs of wear or incipient failure.

In spite of this one must anticipate that faults will occur during the working life of the system and the purpose of the second safety level is to ensure that such faults can be dealt with safely. The reactor is thus equipped with a protection system which is designed to prevent or to deal with all conceivable fault conditions.

The third level is designed to prevent the escape of radioactive materials to the environment should the second level fail.

The safety systems will have built-in redundancy, or duplication of systems, so that if one should fail a second will still supply the essential service. In practice three or four redundant circuits are usually provided for each system, and the reactor is tripped only when two or more circuits call for action simultaneously. The fact that there are several redundant circuits means that each can be tested separately without tripping the reactor. By frequent testing the chance that single circuit will fail on request can be reduced to one in a thousand requests. The chance that two circuits will fail simultaneously is then one in a million and for three, one in a billion. However this only applies if the circuits are entirely independent. If they are not, for example if they are in a common conduit, a faulty third circuit in the same conduit could put both out of action simultaneously. The best way to eliminate such "common mode failures" is by diversity, that is, not only having physically different safety circuits, but ones which work on different principles or measure different things. Thus it is desirable to have at least two separate, different and independent ways of tripping a reactor. Common mode failures have occurred in reactors in the past.

For example in the Hanford N reactor (a graphite-moderated, water-cooled reactor not dissimiliar to the Chernobyl reactor) the reactor operator switched the release magnet of a control rod to mains power to prevent that rod being inserted during a planned trip. Unfortunately as the result of a multiple diode failure all of the rods were connected to the mains so that none could be released when the trip was demanded. An automatic and independent trip system then operated to dump boron-containing balls into the reactor core and shut the reactor down. This is a classic example of an occasion where diversity in the design of systems proved its worth.

Common mode failures played a part in the Brown's Ferry accident and at Three Mile Island. Reactors should also incorporate back-up electrical systems to take care of possible power failures, back-up systems for removal of fission product heat after shut down and an ultimate heat sink to which heat can be transferred under all operating and accident conditions. In addition there should be emergency core-cooling systems and finally a containment system as a last-ditch method of preventing the release of fission products to the environment.

All of these safety systems should be capable of being monitored from the control room which should be constructed so that it will be habitable during the course of the most severe anticipated accident. This will involve the provision of a clean air supply and also the shielding of the control room from any radiation emitted during a possible accident. Finally provision should be made for two Site Emergency Control Rooms each having good communications with the reactor control room. It is necessary to have two of them, because, should a release of fission products occur the wind direction may make a single emergency control room untenable. One emergency control room should be sited upwind of the reactor in the direction of the prevailing wind and the other on a line through the reactor perpendicular to the prevailing wind direction. The site emergency control room will be the contact point with the outside world in the event of a severe accident and will also be the point from which monitors depart to make their radiation surveys so that it must always be accessible and clear of any contaminated area.

When all this has been done and the reactor is as safe as good design can make it, the most essential thing is good training, responsibility and discipline for the operating staff. Training must be continuous and progressive and the one thing that must be guarded against at all cost is complacency. Operating rules must be obeyed and no infringements tolerated. The initial cause of the Three Mile Island accident was a breach of operating instructions and the apparent cause of the Chernobyl accident was an unauthorised experiment done in an unapproved way. Everyone working on a reactor should be highly selected, highly trained and highly paid. The responsibilities of the job demand no less. Airline pilots are paid more than bus drivers not primarily because the job is that much more difficult but because the potentialities for disaster are greater, the requirement for split-second correct decision making is ever present and no lapses of concentration can be tolerated. The same differentials apply to reactor personnel as opposed to those in conventional generating stations.

One of the ever present dangers among nuclear reactor operators is boredom. When a reactor is running normally there is really very little for the operators to do. Yet somehow they must always be kept on the alert for any sign of an impending malfunction and trained to respond instinctively. The psychological demands of such a job are extraordinarily high and should require stringent psychological testing of all applicants, quite apart from their technical qualifications. There is some evidence that such testing is not universal and that the technical qualifications demanded are sometimes barely adequate.

Security

There have been no reports that any terrorist has as yet tried to hijack a reactor or to blackmail a government by the threat to cause a Chernobyl-type disaster to a power reactor unless specified demands are met. It is easy to downplay the dangers of such an eventuality by saying that the terrorist would have to have specialist knowledge and training. Experience has shown that any specialist training can be acquired by committed terrorists with the support of sympathetic and unscrupulous governments. It will be argued that access to the reactors could be controlled, but having to show a pass to a security officer at an entrance gate is not really an effective barrier to a determined band of terrorists. A reporter was recently able to smuggle a dummy revolver through the security network at a Canadian airport, and the airports have had twenty years to devise effective systems. It is obviously not in the public interest to disclose details of operating security systems at nuclear generating stations and without doing so one cannot assess how effective they are. Suffice it to say that this is also an area that demands constant vigilance.

The Probability of Accidents

When all of these criteria have been met how does one forecast the probability of an accident to a reactor? The whole system is so complex that it would seem an impossible task. Nevertheless a number of attempts has been made. The first was a study entitled "Theoretical possibilities and consequences of major accidents in large nuclear power plants" which was published in March 1957. It was a completely arbitrary study and there never was any justification for the assumptions made in it. At this time there was very little actual experience of running power reactors, so the study was indeed theoretical. It considered the case of a 200 Mw(e) reactor, situated 50 km from a city of 1 million people which had an accident resulting in the release of half of its fission products to the atmosphere. In the worst case it was estimated that 3400 deaths and $7 billion property damage might occur. The probability of such an accident occurring was put at between one chance in 100,000 and one chance in 1000 million per reactor year. This means that there is a 63 per cent chance (or slightly better than even) that the accident will occur in 100,000 or 1000 million years, whichever is correct, for a single reactor, and a smaller chance that the accident will occur sooner. If there are 100 such reactors operating the time will be cut down proportionately to 1,000 and 10 million years respectively. These are very small odds indeed, but people are impressed not by the unlikelihood of the event but by the consequences should it ever happen.

The Rasmussen Study

In 1972 the US Atomic Energy Commission sponsored a further comprehensive study to attempt to assess the risks to the general public from accidents to commercial reactors such as were operating at that time. The study concerned itself with light-water reactors and as examples chose a single PWR, the Surry Unit 1 (788 Mw(e)) and a Boiling Water Reactor, Peach Bottom, Unit 2 (1065 Mw(e)). The procedure used to estimate the probability of a rare accident was to consider the chain of events which could lead to the accident and to assign probabilities to each event in the chain which were based on or inferred from previous experience. If the probability of a certain event in this "event tree" is not known from past experience it can be derived from a "fault tree." In this every pathway of faults which leads to the event in question is scrutinised and the probability of each determined so that the probability of the top event can be calculated for use in the event tree.

The great difficulty is in estimating the probability of failures of which there has been little or no experience, although there has been vast experience in the reliability of most of the components of reactors. Several hundred event trees were examined in this study and many of them could be neglected as having insignificant probability of occurrence or trivial consequences. The remaining event trees were then subjected to detailed examination.

The report of the study "An assessment of accident risks in US commercial nuclear power plants" was submitted to the Nuclear Regulatory Commission in 1975 and published as NRC Report WASH-1400 (NUREG-75/014). It is often referred to as the Rasmussen report, after the director of the study, N. C. Rasmussen. The report has five major sections:

1. the identification of accident sequences which could lead to core meltdown and the determination of the probability of each such sequence

2. the radioactive releases from core meltdown for various containment failures

3. the dispersion of the released radioactive materials in the atmosphere

4. the assessment of health effects and property damage

5. the assessment of fatalities and injuries

WASH-1400 gave the probability of an accident causing one early fatality as 1 in 2.5 million reactor years and that of an accident causing 2500 early fatalities as 1 in 1000 million reactor years. From the uncertainties in the data it was considered that the given probabilities might be a factor five too low or too high.

Review of the Rasmussen Report

In 1977 the NRC set up a Risk Assessment Review Group, consisting of a number of independent experts to review WASH-1400. The findings of the Group were accepted by the NRC in January 1979. The Group reported that the Reactor Safety Study represented a considerable advance over previous attempts to estimate the risks of nuclear power plants. It considered that the method used was sound although the data base was not adequate in all respects; the study had not been able to quantify common mode failures and did not place enough emphasis on the possibility of accidents caused by earthquakes, fires and human error. Because of this the review was unable to arrive at a conclusion as to whether the absolute probabilities determined by the Study were too high or too low, but it concluded that the possible errors were greater than suggested in the study. Nevertheless the group found that "despite its shortcomings WASH-1400 provides at this time the most complete picture of accident probabilities associated with nuclear reactors."

In evidence before the Congressional Sub-Committee Dr. Harold Lewis, the Chairman of the Review Group said,

> We concluded that the Reactor Safety Study was a conscientious and honest effort to quantify the risk and consequences of a reactor accident and, although it fell short of the mark, it was still the best single available source of information on these matters. We found that the methodology was sound, but that the implementation of the methodology was imperfect for a number of reasons, among them an inability to adequately quantify common cause failures, an inadequate data base, and inadequate statistical techniques. We were able to find a number of places in which our view was that the treatment understated the risk, and others in which it overstated the risk, and we were not, on balance, able to say which predominated. However it was clear to us that the error bands stated in the report (a factor of five in either direction) greatly under-

> stated the fragility of some of the calculations. Even though the *Washington Post* characterised me a week ago as an 'anti nuclear cult hero,' I have to tell you that I am pro-nuclear. There are many reasons, but I generally feel that the risk is small, and that it is the potential for a large and *unfamiliar* event that so frightens people that the risk tends to be exaggerated. The demonstrated risk of coal mining, the risk of war over oil, the risk of a decline in our economy — all these are worse. Further, of the three issues involved in the nuclear question — proliferation, waste disposal and safety — I regard the first as insoluble but controllable, the second as trivially soluble technically, but as yet unsolved and the third as in much better shape than the other two. Even on the safety issues, my personal view is that the conservative elements in WASH-1400 overbalance the non-conservative ones, and that, in particular, failure to account for constructive human response during the course of an accident is an omission which outweighs all others.

In recent years many of the opponents of nuclear energy have given the impression that the Rasmussen report was in some way discredited by the review group. As has been seen, this is not true. The occurrence of the Three Mile Island accident in 1979 did not vitiate the Rasmussen report in any way. On the number of reactor operating years accumulated by that time, an accident of that degree of severity (there were no immediate casualties) was to have been expected. There is no relevant basis of applying the Rasmussen report to the Chernobyl disaster — it was developed for a different kind of reactor.

Radioactive Wastes

The unique feature of nuclear reactors compared to other energy producers is that their waste is highly radioactive and must not be released to the environment. Each atom of uranium which undergoes fission produces two other atoms which are radioactive and decay through chains which often contain three or four other radioactive atoms. Reactor fuel which has been used for energy-generation is thus very much more radioactive when it comes out of the reactor than it was when it went in. Fuel removed from a reactor therefore has to be heavily shielded, to protect the operator and also cooled to remove the heat generated by radioactive decay of the fission products. Most of the fission products have rather short half lives so that even after one day the radioactivity has substantially decayed. Even at one day's decay, however, there will be nearly a million curies of fission product activity in the fuel inventory for every megawatt (heat) at which the reactor was operating. It is not the short-lived fission products that are the major problem, but the long-lived fraction which will remain radioactive for many years. The problem is simplified if no fuel reprocessing (as, for example, in Canada) is carried out and the used fuel elements are merely stored under water, which provides both cooling and shielding. Some countries (for example Britain and France) reprocess the used reactor fuel to recover the unused uranium 235 and the plutonium which has been formed, for use both in fast reactor and weapons programmes. The chemical reprocessing produces as an additional product highly radioactive acid solutions. These have, to date, been retained in stainless steel tanks which are cooled. The estimated lifetime of the tanks is about 30 years so that a succession of such tanks will have to be built in the future to replace aging and worn out ones. It has long been recognised that methods must be developed for the safe management of these wastes. In 1951 Sir John Cockcroft said, "the disposal of radioactive waste at one time seemed to be one of the major problems of a large-scale atomic energy programme . . . . The problem is to separate by radiochemical methods the fission products which are long lived and to concentrate and store them."

The feasibility of several methods has been demonstrated on a fairly large scale. The two methods likely to be used in the future are (i) drying and calcination and (ii) vitrification. In the first the highly active liquid waste is first heated and dried and the solid residue further heated to drive off any volatile material. The resulting powder may then be stored. The advantage of this method is that a smaller volume of solid waste is obtained and the disadvantage is that the thermal conductivity is low, making the removal of the fission-product-decay-heat difficult. In one method of vitrifying the waste a mixture of borax and silica is added to the calcine at about 600°C and the mixture heated to about 1000°C. When cooled the result is a glassy mass. A tonne of irradiated uranium will result in between 30 and 50 litres of calcine or between 60 and 80 litres of vitrified material. Borosilicate glass has relatively good thermal conductivity and very high resistance to leaching. A single 1000 Mw(e) reactor would produce about 3 cubic metres of highly active solidified waste per year. For ultimate disposal the solidified wastes would either be stored in disused mines or sunk in shafts in geologically stable formations.

An alternative method to vitrification is the incorporation of highly active waste in "synroc," a synthetic rock manufactured from titanium, calcium, zirconium, barium and aluminum oxides. This is claimed to have a lower leach rate than glasses and simulated aging using bombardment by fast neutrons suggest that at a simulated age of 200,000 years there was no marked increase in leaching of caesium. Development of this method is proceeding at the Australian AEC Establishment at Lucas Heights, New South Wales.

There is much concern among the environmental organisations that it is dangerous to bury highly active wastes in this manner and that to do so is merely to pass on insoluble problems to future generations.

Current proposals call for the sealing of vitrified material in 12 mm-thick stainless steel cans, surrounding them with 10 cm lead and 6 mm of titanium and placing them at a depth of 300–1000 metres in a dry and stable geological formation. As Fremlin (1980) says:

> It is difficult to see how they can return to the surface in less than 600–700 years when their activity will have fallen to that of the original uranium ore from whence they came, except that they will not be emitting radon. Even in flowing saline water the titanium layer will last for at least 4000 years and the lead layer for longer still. Finally, from the evidence of the natural reactor at Oklo in Gabon, almost all the fission

products and trans-uranic elements stayed within a metre or so of their source for 1700 million years.

The low and intermediate level wastes occur in greater volume and are less hazardous. There is no reason to suppose that the methods at present being used for their storage will be unsatisfactory in the future.

Chapter Eleven

After Chernobyl: What Now?

The disaster at Chernobyl, for disaster it undoubtedly was, caused thirty-one deaths among the plant workers, disrupted the lives of thousands of people and will undoubtedly cause further deaths in Russia and other countries in the next thirty years, although it will never be possible to link specific deaths directly with the accident. It also created a chill of fear around the northern hemisphere which was intensified by the reporting of the news by the western media. The Soviet government complained, and with some reason, at the sensationalist reporting of the tragedy in the capitalist countries and at the barely concealed satisfaction of the many instant experts who mushroomed out of obscurity and forecast greater and greater disasters to come. The official US reaction was scarcely more praiseworthy with Secretary of State Shultz offering to bet $10 that the casualties were greater than the Soviet government would admit. At the time of the Three Mile Island accident the Russian attitude was not very different and in recent years the Soviet authorities have taken a certain amount of public satisfaction from the chronicle of minor mishaps in western reactors and contrasted them with the impeccable safety record of those in Russia. Even after the Chernobyl accident *Isvestia* reported that in 1979 there were 2310 incidents at 68 American nuclear power plants including the "especially severe incident at Three Mile Island." By 1980 the number of reported incidents at US plants had risen to 3804 and by 1981 it was 4060, more than eleven per day. The purpose of the article was, of course, to point out that while the USSR had had only one "incident" — Chernobyl — the United States has had more than 20,000, quite overlooking the fact that in western democracies all such incidents are reported publicly while in Russia they are not. There was also a great deal of western criticism of the delay in the release of details of the accident to other countries. This was certainly a serious error, not so much on the part of the Soviet government which had no conception of the magnitude of the accident, if indeed they were aware that there had been one, but by the local management at Chernobyl. Their reaction was wrong, but understandable. It is always a natural reaction to hope that an accident is not as bad as you think and that you will be able to retrieve the situation without causing a lot of fuss, but it is a reaction that must be suppressed.

Appropriate response to accidents must be planned in advance. That is not so silly as it sounds. In every reactor control room there should be a list of all conceivable fault conditions and for each one there must be a specified level at which the Site Emergency Organisation must be called into play. It is not at all clear that there was such an organisation in existence at Chernobyl, if it did exist it was apparently ineffective. There was a considerable delay before evacuation was ordered and this implies that all of the resources of the station were marshalled into dealing with the immediate emergency and that no thought was given to the possible consequences outside the station. Valentin Falin, the head of the Novostni press agency said in Bonn (17 May) that Mr. Gorbachev was informed of the accident a few hours after it happened but that he was evidently unaware of the seriousness of it at that stage. Telephone calls were also made to top officials at the ministry and state committee responsible for atomic energy at about 5 am on the Saturday, three and a half hours after the accident began.

Despite these top-level calls, confusion and lack of coordination continued. Firemen battled the fierce flames with their boots sticking in the molten bitumen. Gennady Berdov, deputy interior minister in the Ukraine, coordinated the firefighting, just as he would have any large-scale local fire. Astoundingly, as later admitted by *Tass*, neither Berdov, nor anyone else, ever imagined that there could be any radioactive material escaping from the reactor that would cause the deaths or serious illness of almost everyone present, let alone cause health problems as far away as Finland and England. This implies the absence of an effective site emergency organisation. No one doubts that the firefighters behaved heroically in the accident for which they had never been psychologically trained, or even prepared. The tragedy is that they were apparently thrown into a hopeless struggle and paid for it with their lives.

It is profoundly to be hoped that one of the positive aspects of Chernobyl will be to encourage a great deal more openness between the eastern bloc and the west on questions of reactor safety. There are indications that this may be already occurring. After the initial clampdown on information the Soviet government became very much more open about releasing details of casualties, of the evacuation and even of the punishment of individuals whose performance during the accident had been less than expected. The accident will have other effects on the world's nuclear power programme, both positive and negative. For many people the Chernobyl accident will be the realisation of their worst fears about nuclear power — the "China Syndrome." The word "meltdown" was used freely in media reporting — one headline announced "worse than a hundred meltdowns." In fact, there was no meltdown in spite of the fact that a reactor which was admitted by the Russians as being difficult to control was deliberately mishandled in an insane way. The automatic emergency shut-off system was disconnected, the control rods were withdrawn beyond permissible levels, the automatic control system was disengaged, thus allowing the core to become unstable, the emergency cooling system was disconnected. It is impossible to imagine a more complete recipe for disaster in reactor operation.

If the Russian operators were not aware of the idiosyncracies of their reactor they should have been. Writing in the journal *Atomnaya Energiya* in June 1984, E. V. Kulikov said that "the stability in the power distribution decreases as the fuel burns up" and that new systems had to be added into the design to compensate. One system provides "local automatic control" of the power in individual regions of the reactor and the second system "allows for emergency reductions in power when there are impermissible local power rises in spite of the action of the automatic control system." These were two of the systems disengaged by the operators. It is difficult to imagine such a series of deliberate maloperations being allowed to occur in any western nuclear station, where even the most junior operator is apt to argue if he is told to do something which appears stupid or dangerous. One must suppose that only the autocracy of the Soviet system and the reluctance to question the decision of a superior can have allowed it to happen in Russia.

Effects on Other Nuclear Power Programmes

After their initial outbursts of near-hysterical joy the reaction of the anti-nuclear environmental organisations to the Chernobyl accident has been remarkably muted and it seems that they must have realised that there are very few points of similarity between the situation at Chernobyl and any western nuclear generating station. The RBMK 1000 reactors are almost unique to Russia. The Soviet government has never exported them, even to satellite countries. There is one western reactor that bears a basic similarity to the RBMK reactor (in fact the Russian design may have been copied from it), the N Reactor at Hanford, Washington, which is also a graphite-moderated water-cooled reactor.

There are important differences. The fuel burnup in the RBMK is 18000 megawatt days per tonne while that in the N reactor is 1000. The graphite operating temperature in the RBMK is 700°C while in the N reactor it is 325°C; the N reactor is also considered to be the last of its kind and it is certainly very unlikely that any similar reactors will ever be constructed outside Russia.

The British nuclear power programme uses graphite-moderated rectors but, again, the graphite runs at a much lower temperature than in the RBMK and the reactor is cooled by carbon dioxide rather than by water. It is thus not possible for an RBMK-type accident to occur outside Russia, but that does not mean that there are not valuable lessons to be learned from the Russian experience. There should not a be a single reactor anywhere whose safety regulations are not being closely scrutinised in the light of Chernobyl and, from this point of view, the effect of Chernobyl on the rest of the world will be highly positive.

Emergency Procedures

One must hope also that the emergency procedures of civil authorities everywhere are being carefully reassessed as a result of the Chernobyl disaster. The existing emergency plans for various areas are summarised below.

Britain

In the event of an accident at a British plant posing a hazard to the general public an elaborate emergency planning system would be triggered. Everyone living within 2.5 kilometres downwind of the plant would be evacuated and others living up to 5 km away would be told to stay indoors and to keep their windows and doors closed. Police throughout the country keep a register of people living close to nuclear installations which is updated every year. Town halls would be requisitioned by the government as emergency control centres. For example if an accident were to occur at Dungeness on the south coast of Kent, the town hall at Tenterden, 25 kilometres from the reactor, would become the crisis headquarters. Telephone and telex lines are already installed there and elsewhere. A government-appointed emergency controller would immediately assume virtually dictatorial powers over every aspect of life within a radius of 40 kilometres from the reactor, coordinating the work of the emergency services, hospitals and government departments. The Ministry of Agriculture would ban the consumption of milk and other food produced within the affected area and, if necessary, water supplies would be brought in from outside. Local health authorities already stock potassium iodide tablets which would be widely distributed, especially to children in an attempt to minimise the uptake of radioactive iodine to the thyroid.

According to the emergency document for the Sizewell (Suffolk) station, the measures deal with the situation which could arise if the worst type of pressure circuit rupture occurred and coincided with a fire in part of the reactor. The Central Electricity Generating Board stresses that such an emergency is "extremely unlikely" but that its staff are highly trained to deal with it efficiently and without panic. The trigger for a site alert would be a rapid rise of radiation levels. Station staff would be issued with gas masks, protective clothing and potassium iodide tablets to provide first-line protection against contamination. They would seek cover indoors at pre-selected emergency assembly points. Once the emergency alarm had been sounded throughout the plant an on-duty health physics officer would assess the hazard to the public.

A duty rota is maintained at the station with telephone numbers to ensure that key staff can be contacted 24 hours a day within minutes of an accident. It is realised that it might take 30 minutes for them to arrive at the site, but other staff have been trained to deputise. The police and emergency services would be warned, together with the Atomic Energy Authority, the Nuclear Installations Inspectorate and neighbouring nuclear power stations. Four control rooms at the site would come into action immediately. The rooms are already equipped with maps, plans, technical information and VHF radio and telephones. Advice to police to begin dispensing potassium iodide tablets and organising evacuation of the affected areas would be based on the results of air sampling, deposition measurements and radiation levels. Staff with a fleet of radio cars would continually monitor radiation levels, air and ground contamination in the surrounding area.

Every nuclear installation in the country is required by law to rehearse its emergency plans fully at least once a year. However a recently planned site emergency exercise at Bradwell (Essex) was called off by government intervention because "local groups feared that the exercise would create anxiety about nuclear power instead of reassuring local residents." This is such a fatuous argument that it is almost incredible. How can one reassure local residents without convincing them by such exercises that the reactor staff is prepared for any eventuality? In the face of such an ill-advised decision one can have some sympathy with the characteristically destructive criticism of Friends of the Earth in the person of Mr. Stuart Boyle: "The plans really are a lot of hogwash. The British nuclear industry is so complacent that it refuses even to plan for an accident only half as big as Chernobyl."

USA

In the United States reactors for generating electricity are operated by local power companies under a licence from the Nuclear Regulatory Commission which will not issue it without an approved emergency-planning scheme. According to the NRC the nuclear power centre at Indian Point, New York would probably be the most hazardous site at which to have a major release of radioactive material.

The Commission has concluded, however, (after a six-year enquiry) that because of their design and safety equipment, the risk they pose is only a small fraction of those incurred by the surrounding population from such everyday risks as automobile accidents and fire. According to a probabilistic study carried out by Con Edison and the State Power Authority, each of whom operates one of the Indian Point reactors, the likelihood of an accident involving a meltdown at the plant is once in 9,100 years. Of all the accidents studied, only about 1 in 1000 meltdowns would lead to a release of radioactive materials leading to early deaths and about one meltdown in 100 would release enough radioactive material to cause immediate illnesses. About half the meltdowns could release enough radiation to cause latent illnesses like cancer.

Combining the risk of meltdown with the risk of a large release results in an estimate that an accident causing immediate deaths could be expected once in 250 million years. The safety review followed a petition in 1979 by the Union of Concerned Scientists which has attacked the methodology of the study. The technique involves listing all the equipment which would have to fail to cause an accident or a release of radioactive material and then computing the probability that enough of that equipment would fail in sequence. Critics say that the estimates cover only those sequences of malfunctions which can be anticipated and that the probability of some categories of causes, like sabotage, can be difficult to estimate.

The Indian Point reactors have been among the most controversial in the USA because of the population density around them and their proximity to New York City. More than a quarter of a million people live within 10 miles of the site which is on the Hudson River 35 miles north of Manhattan. The population within 50 miles of Indian Point is 17 million. Since the Three Mile Island accident the NRC has required reactor operators to present plans for evacuating or sheltering people living within ten miles of a reactor and to make provision for testing food and water up to 50 miles away. As a result of the Indian Point Safety Review, the NRC decided that, with minor modifications, the operation of the two reactors did not constitute an undue risk to public health and safety.

Canada

Most of Canada's nuclear electricity is generated in Ontario and the provincial cabinet minister responsible for nuclear emergency planning is the Solicitor General. His final draft report was approved by the Ontario cabinet in June 1986. The report is based on the assurances of Ontario Hydro that no accident at any of the nuclear stations in Ontario would be as extreme as that at Chernobyl. The report advocates the setting up of Evacuee Reception Centres for people evacuated from their homes (which would also take care of children removed from schools until their parents arrived to take care of them). It makes the point that in certain circumstances sheltering, followed by evacuation, may provide the optimum solution, if, for example, a large emission occurred with insufficient time to evacuate people before the arrival of the plume, or "if it is safer to wait out adverse but transitory environmental conditions."

The report comes to some commonsensical, but nevertheless valid, conclusions. It states, for example, that "Emergency planners and managers must take into account the likelihood that, once the emergency is known to have occurred, many people will leave the area spontaneously" and also that, while selective evacuation of pregnant women, young children, seriously ill or bedridden patients and the handicapped may make sense, such an evacuation would be likely to trigger a general but uncontrolled exodus from the area. The report also proposes a series of government groups to disseminate information to the public as well as a media monitoring organisation to assess public reaction and a "rumour control desk."

The report raises a number of questions — What would the authorities do if people outside a cordoned-off area were to try to reach their families inside the area? Would it be better to risk leaving or to stay indoors until the danger has subsided? The report advocates the taking of potassium iodide tablets to block uptake of radioactive iodine to the thyroid. As mentioned in Chapter Three this is a controversial question. It states that the tablets must be taken within 12 hours or they are virtually useless and points out that their use would require the timely distribution of the tablets to the affected population with the appropriate instructions.

About 40 per cent of Ontario's electricity comes from nuclear reactors and the Nuclear Emergency Plan lists seven "hazard" zones which include the areas near the US border which are considered to be at risk in the event of an accident to an American reactor. The Pickering plant 40 kilometres from Toronto has seven reactors and 100,000 people living within 10 kilometres.

In the event of an accident involving a serious release of radioactive material the Emergency Planning Organisation would advise the Premier to declare an emergency. The Premier would then designate the Solicitor General to take overall command of the emergency organisation at the Ontario Provincial Police headquarters near the Toronto waterfront. There technical advisers would meet representatives of the affected area's police forces and fire departments and, after studying reports from teams monitoring radiation levels around the plant, would advise on the action to be taken.

The Emergency Planning Organisation is not without experience as it masterminded the highly successful and fatality-free evacuation of 250,000 residents from their homes after the Mississauga train derailment of November 1979. The EPO coordinator, Kenneth Reeves, stressed that the successful operation demonstrated the crucial importance of good communications during an emergency. He pointed out that the officials at Three Mile Island had failed to keep the press and public informed and suggested that, if they had, the confusion and the large voluntary evacuation would not have occurred.

The Future

After the Chernobyl accident, a collection of consumer, environmental and scientific groups known as the Coalition of Environmental/Safe Energy Organisations called for a complete phase-out of nuclear power plants in the USA. In Pennsylvania protesters vowed to increase efforts to prevent the reopening of the second reactor (which was not involved in the accident) at Three Mile Island. In New Hampshire and Long Island antinuclear forces stepped up their campaigns against the licensing of the Seabrook and Shoreham plants. The Union of Concerned Scientists said, "The accident at Chernobyl makes it clear that nuclear power is inherently dangerous." So is every other sort of power and all of the evidence tends to show that nuclear power is less dangerous than the others.

In spite of this, it seems highly probable that the Chernobyl accident, disastrous though it was, will have much less effect on the future of nuclear power than the Three Mile Island accident which did not cause a single casualty. It is widely recognised that the RBMK-type reactor, with its red-hot graphite interspersed with tubes full of boiling water and steam under pressure, is unique in the world and an accident of this type is impossible outside the Soviet Union.

What then will be the future course of nuclear power? The Paris-based Nuclear Energy Agency expects that by the end of this century nuclear power generation will increase in the industrial world by about 85 per cent and that by then more than a quarter of the electricity produced by the 24 member-countries of the Organisation for Economic Cooperation and Development (OECD) will come from nuclear reactors. Even if no new plants were ordered total nuclear power station capacity in OECD countries would rise by 50 per cent because, once built, nuclear stations tend to be run flat out for their operational lives leaving the more expensive (to run) coal-fired and oil-fired stations to deal with intermittent demand at peak load periods. For this reason the proportion of power produced from nuclear energy is expected to increase faster than the installed capacity. France is expected to maintain a decisive lead up to the end of the century by which time it will be producing 78 per cent of its electricity from nuclear power compared with about 65 per cent at present. Although the United States is expected to have about half as many plants again as France by the end of the century, it will only be generating about 18 per cent of its energy from them.

Future Reactors

Will any new reactors ordered in the next ten years be significantly different from those operating now? It seems as though they may well be very different. The greatest change from the start of nuclear power to the present has been in the size of the stations. The Russians claim that the first operating nuclear power station started at Obninsk in 1954. It was a 5Mw(e) station and was closely followed by the UK's Calder Hall which had a 50 Mw(e) output. The early Magnox stations in the UK had outputs of about 275 Mw(e), the AGR's 620 Mw(e) and the PWR's 1050 Mw(e). It seems to be a clear case of "bigger is better."

Current thinking seems to be that rather "small is beautiful" and designs are appearing (Lester, 1986) for much smaller and inherently safe reactors. In 1956 a group of scientists was assembled by a new division of the General Dynamics Company called General Atomic. They were given the task of designing a commercial nuclear reactor and Edward Teller (the inventor of the hydrogen bomb) said he wanted it "safe enough for a party of school children to play with."

The result was TRIGA which was completed three years later. At the dedication ceremony the Nobel Prize-winning physicist Niels Bohr did an unheard-of thing. He pressed a switch which pulled out all of the control rods at a very high speed. It should have caused a catastrophic accident. For a few milliseconds the output of the reactor soared. Then it brought itself under control. TRIGA was inherently safe.

The largest TRIGA built was a 10 Mw reactor, too small for widespread use as a power generator. A small high-temperature gas-cooled reactor recently developed in Germany has a "pebble bed" core in which the fuel is embedded in small graphite spheres about the size of a tennis ball. One drops out of the core when another drops in. The fuel balls are capable of withstanding temperatures of 1600°C without releasing fission products. The small size and large surface to volume ratio of the reactor ensure that the core temperature would not get to that value even if the reactor lost all its helium coolant while operating at full power. After an initial rise of temperature the chain reaction would stop automatically, the fission-product decay would still produce heat, but the loss of heat from the walls of the reactor vessel would be sufficient to stabilise the temperature at a safe level. Because the small size is crucial to the safety feature of the design, such a reactor could not produce more than 100 Mw(e) but a generating station could, of course, have several reactors.

A Swedish reactor design, PIUS, (Process Inherently Ultimately Safe) is a light-water reactor in which the core, primary cooling system and steam generators are immersed in a large pool of borated water within a pre-stressed concrete pressure vessel. The pool and the primary cooling system are hydraulically connected but, under normal conditions, the pressure developed by the coolant circulating pumps is just sufficient to prevent any pool water from entering the core. Any non-trivial disturbance of the cooling system would alter this balance and allow borated water to enter the core. Boron is such an efficient absorber of neutrons that the chain reaction would shut down, and the water in the pool would carry off the fission-product-decay heat. These events would be set in motion without the intervention of any operator or electronic control circuit.

A third proposal is for the construction of small (100 Mw(e)) fast-breeder reactors. Their small size would enable them also to dissipate heat harmlessly from the walls of the reactor vessel after an emergency. The advantage of the fast-breeder option is the opportunity of generating more fuel. The disadvantage is the necessity of a fuel-processing system to separate it and the concern that fuel-reprocessing and recycling of plutonium would encourage the proliferation of nuclear weapons.

One of the greatest advances in fast-breeder safety came a month before the Chernobyl accident when the US Department of Energy did its best to destroy a fast-breeder reactor at the National Reactor Testing Station at Idaho Falls. The tests were intended to try out a new design in which the core sits in a large pool of sodium kept at atmospheric pressure. The reactor fuel was a special metallic fuel which is a good conductor of heat, is difficult to rupture and expands when heated so that the nuclear reaction slows down. With the reactor working at full power the cooling system was cut off, simulating a loss of electricity to the station. When the fuel and sodium heated up the reactor automatically shut itself off. The core did not overheat and there was no damage to the fuel. In the second test the secondary cooling system was shut off as well. The reactor shut itself off as before without any intervention from the operator or any electromechanical devices.

It has already been shown that the resources of fissile material on the earth are not sufficient to maintain a nuclear energy programme for a world using energy at the western industrialised nations' rate for more than a few tens of years. Thus nuclear energy will occupy no more than a ''blip'' on the earth's energy-producing history. This is perhaps a pity for there seems to be little doubt that a given quantity of energy will be generated with less cost in human suffering from nuclear resources than from any other method in the earth's history. The objections of the anti-nuclear groups are more emotional than rational. They are also rather hypocritical. There are undoubtedly risks from nuclear power, even though they are less than from any other method of generating energy. The novel feature of nuclear power is that, for the first time in history, the risks are divided among the users of electricity rather than concentrated on a small section of society. The nuclear protesters do not appear to use less electricity than the rest of society but they seem to object to paying their share of the social cost.

References

Chapter One

Beattie, J. R. and Pamela M. Bryant. Assessment of environmental hazards from reactor fission product release. United Kingdom Atomic Energy Authority, AHSB (S) R135. London, Her Majesty's Stationery Office, 1970.

Reactor explodes amid Soviet silence. *Science* 232, p. 814, 16 May 1986.

The world's worst nuclear accident. *New Scientist* 1506, p. 17, 1 May 1986.

Interim report on fallout situation in Finland from 26 April to 4 May 1986. Helsinki, Finnish Centre for Radiation and Nuclear Safety (STUK), STUK-B-VALO 44, May 1986.

Second interim report, Radiation situation in Finland from 5 to 16 May 1986. Helsinki, Finnish Centre for Radiation and Nuclear Safety (STUK), STUK-B-VALO 45, May 1986.

The hazards to man of nuclear and allied radiations: A second report to the Medical Research Council. London, Her Majesty's Stationery Office, Cmnd. 1225, 1960.

Chapter Two

The accident at Chernobyl AES and its consequences. State Committee for Using the Atomic Energy of the USSR, August 1986.

Chapter Three

Megaw, W. J. The penetration of iodine into buildings. *International Journal of Air and Water Pollution* 6, pp. 121–28, 1962.

Chapter Four

The implications of the new data on the releases from Sellafield in the 1950s for the conclusions of the Report of the Investigation of the Possible Increased Incidence of Cancer in West Cumbria. Committee on Medical Aspects of Radiation in the Environment (COMARE), First Report. London, Her Majesty's Stationery Office, 1986.

Court-Brown, W. M., Doll, R., and A. B. Hill. The incidence of leukemia following exposure to diagnostic radiation *in utero*. *British Medical Journal* 2, p. 1539, 1960.

Hall, E. J. The biological effects of low doses of ionising radiations. Gordon Richards Memorial Lecture to Canadian College of Physicists in Medicine, 1982.

Hawkes, Nigel et al. *The Worst Accident in the World, Chernobyl: The End of the Nuclear Dream.* London, Pan Books and William Heineman, 1986.

Hutchinson, G. B., MacMahon, B., Jablon, S., and C. E. Land. Review of report by Mancuso, Stewart, and Kneale of radiation exposure of Hanford workers. *Health Physics* 37, p. 207, 1979.

Investigation of the possible increased incidence of cancer in West Cumbria: Report of the Independent Advisory Group. London, Her Majesty's Stationery Office, 1984.

Jablon, S. and K. Kato. Childhood cancer in relation to prenatal exposure to atomic-bomb radiation. *Lancet* 2, p. 1000, 1970.

Kneale, G. W., Stewart, A., and T. F. Mancuso. Re-analysis of data relating to the Hanford study of the cancer risks of radiation workers. *Late Biological Effects of Radiation*, Vol. 1, p. 386. Vienna, International Atomic Energy Agency (IAEA), 1978.

Mancuso, T. F., Stewart, A., and G. W. Kneale. Radiation exposures of Hanford workers dying from cancer and other causes. *Health Physics* 33, p. 369, 1977.

Najarian, T., and T. Colton. Mortality from leukemia and cancer in shipyard nuclear workers. *Lancet* 1, p. 1018, 1978.

Stewart, A., Webb, J., and D. A. Hewitt. A survey of childhood malignancies. *British Medical Journal* 1, p. 1495, 1958.

Stewart, A. and G. W. Kneale. Radiation dose effects in relation to obstetric X-rays and childhood cancers. *Lancet* 1, p. 1185, 1970.

Swedjemark, G. A. Radon in dwellings in Sweden. *The Natural Radiation Environment III* (Gesell, T. F. and W. M. Lowder, eds.), Vol. 2, p. 1237, DOE Symposium Series 51, U.S. Department of Energy, 1978.

Chapter Five

Aaltonen, H. et al. First and second interim reports, Radiation Situation in Finland 26 April to 16 May 1986. Finnish Centre for Radiation and Nuclear Safety (STUK), STUK-B-VALO 44 and 45, May 1986.

Bennerstedt, T., Holmberg, M., and L. Lindborg. Chernobyl fallout: Measurements and consequences. *SSI Report* 86, p. 10, Swedish National Institute of Radiation Protection, July 1986.

Roed, Jorn. Dry deposition in urban areas and reduction in inhalation dose by staying indoors during the Chernobyl accident. Paper presented at MARIA meeting, Brussels, June 1986.

The reactor accident at Chernobyl, Radioactivity measurements in Demark, First, second and third reports. Riso, National Agency of Environmental Protection, May, 1986.

Radiological consequences in Italy of the Chernobyl accident. ENEA-DISP, DOC/DISP (86), 1, Rome, June 1986.

Chapter Eight

Chamberlain, A. C. Emission of fission products and other activites during the accident to Windscale Pile No. 1 in October 1957. UK Atomic Energy Research Establishment (AERE), AERE M3194, 1981.

Crick, M. J. and G. S. Linsley. An assessment of the radiological impact of the Windscale reactor fire, October 1957. National Radiological Protection Board (NRPB), NRPB-R135. London, Her Majesty's Stationery Office, 1982.

Loutit, J. F., Marley, W. G., and R. S. Russell. The nuclear reactor accident at Windscale — 1957: Environmental aspects. Appendix H to the *Hazards to Man of Nuclear and Allied Radiations*, Cmnd. 1225. London, Her Majesty's Stationery Office, 1960.

Patterson, W. C. *Nuclear Power*, second ed., Harmondsworth, England, Penguin Books, 1983.

SL—1 Accident, Investigation Board Report. Congress of the United States, 70593 0, US Government Printing Office, 1961.

Chapter Nine

Thompson, W. B. Prospects for energy. In *Prospects for Man*, ed. W. J. Megaw. Centre for Research on Environmental Quality, York University, Toronto, 1977.

Chapter Ten

Branagan, E. F. and R. L. Gotchy. Health effects of uranium mining and milling for commercial nuclear power. Chapter 9 in *Health Implications of New Energy Technologies*, Ann Arbor, Rom and Archer, 1981.

Dunster, H.J. The assessment of the risks of energy. Forty-ninth Melchett Lecture of the Institute of Energy, 26 November 1981 (subsequently published in Energy World, 1982.

Ferguson, R. A. D. Comparative risks of electricity generating fuel systems in the UK. London, Peter Pereguinus, 1981.

Fremlin, J. H. Risk relativities. *Atom* 283, May 1980.

Lewis, H. W. Risk Assessment Review Group Report to US Nuclear Regulatory Commission. NUREG/CR-0400 Washington DC, 1978.

Muller et al. Standard mortality data for Ontario uranium miners. Proceedings of the International Conference on Occupational Radiation Safety in Mining, Vol. 1, p. 335, Toronto, 1984.

Rasmussen, N. C. Reactor safety study: An assessment of accident risks in US commercial nuclear power plants. NRC Washington, Report WASH-1400, 1975.

Theoretical possibilities and consequences of major accidents in large nuclear power plants. AEC document, WASH-740, 1957.

Chapter Eleven

Lester, R. K. Rethinking nuclear power. *Scientific American* 234, 3 March 1986.

Index

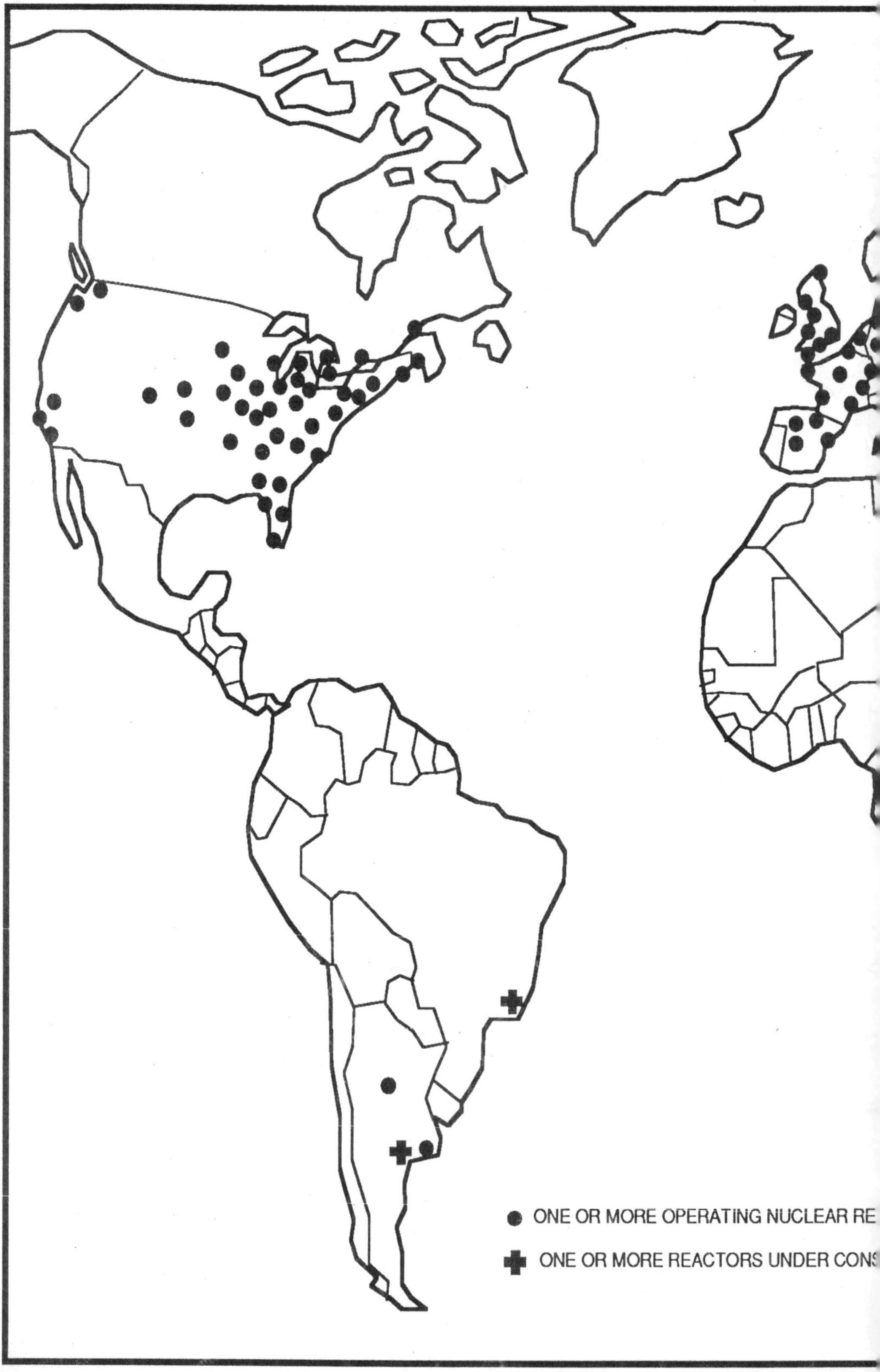
ONE OR MORE OPERATING NUCLEAR RE
ONE OR MORE REACTORS UNDER CONS